REINFORCED CONCRETE
PRELIMINARY DESIGN
FOR ARCHITECTS AND BUILDERS

REINFORCED CONCRETE

PRELIMINARY DESIGN
FOR ARCHITECTS AND BUILDERS

R. E. SHAEFFER

Professor Emeritus of Architecture
Florida A&M University

Consulting Structural Engineer
Tallahassee, Florida

Textbooks, Ltd.
Tallahassee, FL

REINFORCED CONCRETE
Preliminary Design for Architects and Builders, 2nd Edition

Disclaimer

The procedures, equations, tables, charts and graphs in this text are intended for preliminary design only. Any results obtained by using these items cannot be used in the actual design and construction of a building. The author and the publisher accept no responsibility for any consequences resulting from failing to adhere to the above.

ISBN 0-9770795-0-3

The text was set by Jordan Pugh.
The cover was designed by William A. Hees.

678910

Library of Congress Cataloging-in-Publication Data
Pending

Shaeffer, R. E.
    Reinforced concrete: preliminary design for architects and builders, 2nd edition/ R. E. Shaeffer.
    Includes index.
    ISBN 0-9770795-0-3
    1. Reinforced concrete construction.   I. Title
TA683.S533   2006
624.1'8341–dc20

About the Author

R. E. Shaeffer is a registered professional engineer and a Professor Emeritus of Architecture at Florida Agricultural and Mechanical University. He received a B.S. in Building Science from Rensselaer Polytechnic Institute in 1961 and an M.S. in Architectural Engineering from Iowa State University in 1963. He is an experienced structures teacher and the recipient of numerous teaching awards. He has served as a full-time faculty member at several universities including Harvard, Yale, and Cornell.

Professor Shaeffer has authored nine textbooks and more than forty technical papers and currently serves on committees in the American Society of Civil Engineers and the International Association for Shell and Spatial Structures. He has been an invited speaker at ten universities and numerous professional associations.

He is a structural consultant in Tallahassee, Florida and an avid sports car enthusiast.

This effort is dedicated to my family,
and to Roland Hummel, a teacher.

Contents

CHAPTER 1

HISTORY OF CONCRETE BUILDING CONSTRUCTION

CHAPTER 2

CHARACTERISTICS OF CONCRETE AND REINFORCING

CHAPTER 3

REINFORCED CONCRETE BUILDING SYSTEMS

Foreword

R. E. Shaeffer and I have had parallel experiences treading with one foot in architecture and the other in engineering continuously through our careers in education and professional practice. Although this is a slightly perilous journey in terms of maintaining clear values and directions, it does help in the holding of a broadened view of the general field of design of structures for buildings.

Buildings are exceedingly complex and tend to become ever more so over time. Major design concerns now prevalent were never brought up in education or practice only a few decades ago. It is not possible to remain static with one's awareness and general knowledge in terms of design concepts, criteria, procedures, or relative values. What becomes critical in this ever more complex, ever changing situation is to develop and hold onto a firm body of basic understanding of fundamental relationships and ideas.

For both design professionals and educators, however, the real difficulty is to maintain some grasp of the whole view of things and yet be able to whittle down to the simple essence of any given design problem to see the whole and yet be able to make a simple statement that sums it up clearly; to achieve the brilliant clarity and simplicity of the Sermon on the Mount, Hamlet's soliloquy, or the Gettysburg address.

Ron Shaeffer and I eternally strive for the simple view; perhaps a Quixotic search, but, nevertheless, an honorable and significant one in my opinion. We also strive to see design issues from both close up and afar, from the broad vision of the real whole problem as well as from the focused one that zeros in on the specific problem at hand. Having the broad experiences of working in architecture, engineering, and actual construction work, and some involvements in teaching, professional practice, and writing is a valuable asset and helps; but it doesn't guarantee the development of wisdom or a balanced view.

I believe Ron Shaeffer is one of the few truly qualified people around today who can bring a talent for simple viewing and considerable writing skill to the task of explaining significant design issues and procedures for building structures. This book is a notable achievement in that regard, and I recommend it enthusiastically to those readers and students of structural design who truly want to develop a sense of the reality and essence of design work. Concrete is an ancient building material and is presently emerging as a major area of technological development of high sophistication and diversity. Its diversity, permanence, and potential for form manipulation make it attractive to architects, and its structural complexity as a composite material makes it an exciting challenge to engineers. What has been done with concrete to date is history; its future is surely full of many surprises and more realizations as yet not imagined.

It is imperative to the designer's education to develop a simple understanding of the structural nature and responses of concrete and the general problems of producing concrete structures. Added to a well-developed, broad view of the purpose of structures within the whole problem of building design, this book is an excellent background for entering the field of design of concrete structures for buildings. *Reinforced Concrete* represents an ideal opportunity for launching such a learning experience.

James Ambrose
Professor of Architecture
University of Southern California

Preface

*to the 1st edition,
published by
McGraw-Hill, Inc.
1992*

This basic text has been written for students of architecture and building construction and is appropriate for use in certain architectural engineering and engineering technology courses, especially if supplemented by the American Concrete Institute (ACI) Code. Additionally, it can serve as a desk reference for practicing architects and builders who need to review fundamental principles from time to time.

The approach to the subject provides considerably more explanation and background than is available in more simplistic textbooks without the rigor usually found in books for reinforced concrete design courses in structural engineering programs. Intended for use by individuals doing structural planning and schematic or preliminary design work rather than by those responsible for detailed structural designs required for contract documents, the text contains more than the usual amount of material on historical development, system description, and structural appropriateness, and less material on thorough classical analysis techniques and ACI Code-controlled detailing. In the interest of brevity and clarity, emphasis has been placed primarily on developing an understanding of the behavior of concrete elements and systems.

Introductory chapters on foundation systems and prestressed concrete principles have been included, even though these topics are not covered in most texts on reinforced concrete. It is likely that most architecture teachers will not justify requiring their students to purchase separate texts on these subjects, but would nevertheless like to address these topics in a course dealing with concrete structures.

Chapter 16 is an assembly or recapitulation of many of the procedures discussed in the previous chapters, but all pertain to the same building structure. It emphasizes the effects of continuity (so influential on the behavior of most concrete systems) and how loads "travel" or are transmitted from one building element to the next until they reach the ground. Chapter 16 also points out the role of preliminary structural design in space planning and organization.

The Appendix contains graphs that are useful as calculation aids for the preliminary or approximate sizing of beams and columns in the typical reinforced concrete building frame rather than for final structural design, which in most cases must be accomplished by a registered professional engineer. Chapter 4 provides some approximate techniques for estimating maximum moments for preliminary design purposes.

In writing *Reinforced Concrete*, I have assumed that the reader has a background in materials and methods of construction from college-level course work or individual experience. A rudimentary knowledge of indeterminate structures and moment diagrams for frames has also been assumed.

A number of numerical examples are included in the text, and problems for the reader to work occur at the end of appropriate chapters. It is likely, however, that most

teachers will want to supplement these with examples and problems from their own experience. I wish to express my gratitude to several individuals who provided assistance during the development of the original manuscript. My thanks also go to Patsy Harms of the Portland Cement Association for her assistance in locating photographs, to Kathe Gentile who patiently typed from disorganized notes, to Rick Sconyers for helping with the many illustrations, and last *but not least*, to Vicky Newcomb for her invaluable editorial suggestions and for the development of all the problems for students to work. David Billington, Princeton University, and Tony Nanni, Penn State University, deserve my appreciation for their helpful suggestions on this manuscript during the review stage.

I am grateful for the many helpful suggestions made by my editor at McGraw-Hill, Margery Luhrs, who was a delightful cohort.

Finally I wish to thank readers in advance for informing me of errors they find and for making suggestions for improvement in content or approach.

R. E. Shaeffer

Preface

to the 2nd edition,
published by
Textbooks, Ltd.
2006

Most of the topics addressed in the Preface to the First edition apply to this edition. The entire text has been edited to bring the book into compliance with the 2005 version of the ACI (American Concrete Institute) publication entitled, *Building Code Requirements for Structural Reinforced Concrete and Commentary*. This publication is better known as ACI 318, after the number of the ACI committee that produced it. The most significant recent changes have been to the load factors. They are now 1.2 for dead loads and 1.6 for live loads, which brings the factors in line with those specified in the ASCE (American Society of Civil Engineers) publication, *Minimum Design Loads for Buildings and other Structures,* better known as ASCE 7. The capacity reduction (phi) factors have also been modified so that in most cases the member capacity has decreased. The resulting effect is such that for most members, there will be a very small reduction in size or none at all. The most significant change involves beams and slabs, which in all cases will be slightly smaller and/or use less steel. All of the examples and problems have been modified to reflect these new provisions.

Once again, I am indebted to several individuals who assisted me with this project. My thanks goes to Ed Alsamsam of the Portland Cement Association, who helped me get material on tall buildings, and to Tony Felder of the Concrete Reinforcing Steel Institute for his assistance with new steel designations.

Most of all, I am very appreciative of the work done by Jordan Pugh, who worked tirelessly to meet an unrealistic deadline. She had the overwhelming task of converting the book to an electronic format, while accommodating the many changes to the text and figures. She had to re-compose all the pages so that a two-column format was preserved while many of the columns changed in length. It sounds like this would be easy using computer software, but it is not when the paragraphic location of the many equations and figures cannot be changed. She was ably assisted with the figures by her father and my colleague, Professor Tom Pugh.

Finally, as I always do, I plead for readers to notify me of any errors they find. I am *positive* there are some in this edition. I would really appreciate your calling them to my attention, so they can be corrected in the second printing. I am always amazed that no one bothers to tell the author of errors they find. Please also make suggestions for improvement in content or approach.
Thank you!

R. E. Shaeffer
Fax: 850.893.8791

1

HISTORY OF CONCRETE BUILDING CONSTRUC-TION

1.1 EARLY CONCRETE

Much has been written about the numerous significant buildings of the Roman Empire constructed using "concrete" as the primary structural material. Many researchers believe that the first use of a truly cementitious binding agent (as opposed to the ordinary lime commonly used in ancient mortars) occurred in southern Italy in about the second century B.C. A special type of volcanic sand called *pozzuolana*, first found near Pozzuoli in the bay of Naples, was used extensively by the Romans in their cement. It is certain that to build the Porticus Aemelia, a large warehouse constructed in 193 B.C., pozzuolana was used to bind stones together to make "concrete." This unusual sand reacts chemically with lime and water to solidify into a rocklike mass, even when fully submerged. The Romans used it for bridges, docks, storm drains, and aqueducts as well as for buildings.

Roman concrete bears little resemblance to modern Portland cement concrete. It was never in a plastic state that could flow into a *mold* or a construction of formwork. Indeed, there is no clear dividing line between what could be called the first *concrete* and what might be more correctly termed *cemented rubble*. Roman concrete was constructed in layers by packing mortar by hand in and around stones of various sizes. This assembly was faced with clay bricks on both sides, unless it was below grade, and in the case of walls the wythes of bricks served as forms for the "concrete."[1] It is known that the bricks had little structural value and were used to facilitate construction and as surface decoration. There is little doubt that the pozzuolanic material made this type of construction possible, as it was used throughout the Rome/Naples area but is not seen in Northern Italy nor elsewhere in the Roman Empire.

Figure 1.1   The Pantheon

Most public buildings, including the Pantheon, and fashionable residences in Rome used brick faced concrete construction for walls and vaults. The domed Pantheon, constructed in the second century A.D., is certainly one of the structural masterpieces of all time (Figure 1.1). It is a highly sophisticated structure with many weight-reducing voids, niches, and small vaulted spaces. The builders of the Pantheon knew enough to use very heavy aggregates at the ground level and ones of decreasing density higher up in the walls and in the dome itself in order to reduce the weight to be carried. The Pantheon's clear span of 142 ft dwarfed previous spans and created nothing less than an architectural revolution in terms of the way interior space was perceived.[2]

Probably due to the lack of availability of similar pozzuolans throughout the world, this type of concrete was not used elsewhere and stone and brick masonry continued to be the dominant construction materials for most of the world's significant buildings for many centuries. A type of concrete was first seen again in eighteenth-century France, where stuccoed rubble made to emulate true masonry became fashionable. François Cointeraux, a mason in Lyon, searched for an economical means of making fireproof walls by using cementitious mortar in combination with the very ancient *pisé* or "rammed earth" construction technique.[3] Pisé calls for the use of timber formwork to contain the clay or mud while it is being compacted, but the use of new and stronger cements made the compacting process unnecessary. In 1824 Joseph Aspdin, an English mason, patented an improved cement which he called Portland cement because it resembled a natural stone quarried on the nearby Isle of Portland. It is generally believed that Aspdin was the first to use high temperatures to heat alumina and silica materials to the point of vitrification, which resulted in fusion. Cement is still made this way today.

During the nineteenth century concrete was used for many buildings in Europe, often of an industrial nature, as this "new" material did not have the social acceptability of stone or brick.

1.2   THE USE OF REINFORCING

Disagreement exists among researchers as to the first real use of reinforcing in concrete. More often than not, the construction of several small rowboats by Jean-Louis Lambot in the early 1850s is cited as the first successful example. Mr. Lambot, a gentleman farmer in southern France, reinforced his boats with iron bars and wire mesh. He had some plans for using this material in building construction because he applied for a patent in France and Belgium in 1856, describing concrete as follows:[4]

> An Improved Building Material to be used as a Substitute for Wood in Naval and Architectural Constructions and also for Domestic Purposes where Dampness is to be Avoided.

In 1854 a plasterer, William B. Wilkinson of Newcastle-upon-Tyne, erected a small two-story servant's cottage, reinforcing the concrete floor and roof with iron bars and wire rope, and took out a patent on this type of construction in England.[5] He built several such structures and is properly credited with constructing the first reinforced concrete building.

In 1867 Joseph Monier, a French gardener, took out a patent on some reinforced garden tubs and later patented some reinforced beams and posts used for guardrails for roads and railways. It was subsequently shown that Monier never understood, as Wilkinson had, the need for the reinforcing to be near the tensile side of a beam.

The first widespread use of Portland cement concrete in buildings occurred under the direction of the French builder, François Coignet. He built several large houses of concrete in England and France in the period 1850-1880, at first using iron rods in the floors to keep the walls from spreading, but later using the rods as flexural elements.[6]

The first landmark building (Figure 1.2) in reinforced concrete was built by an American mechanical engineer, William E. Ward, in 1871-1875. The house stands today in Port Chester, NY. It is well-known because of the diligence with which Mr. Ward conducted all

FIGURE 1.2   The William E. Ward House. (Robert Mook) [Portland Cement Association]

of his business, researching and documenting everything. He desired a concrete house because his wife was terribly afraid of fire and commissioned architect Robert Mook for the design in 1870. Like Coignet's buildings, it was made to resemble masonry to be socially acceptable. Mr. Ward handled all technical and construction issues himself, conducting long-term load tests and other experiments. He used the French word for concrete, *beton*, and in 1883 delivered a paper on the house to the American Society of Mechanical Engineers entitled "Beton in Combination with Iron As a Building Material." His audience, by definition, was far more interested in the unique water supply and heating systems, which he had designed, than in reinforced concrete.

In 1879 G. A. Wayss, a German builder, bought the patent rights to Monier's system and pioneered reinforced concrete construction in Germany and Austria, promoting the Wayss-Monier system.[7] (Many of these buildings were built in France as well.)

1.3   MONOLITHIC FRAME CONSTRUCTION

The late nineteenth century saw the parallel development of reinforced concrete frame construction by G. A. Wayss in Germany/Austria, by Ernest L. Ransome in the United States, and by François Hennebique in France.

In the 1870s Ernest L. Ransome was managing a successful stone company (producing concrete blocks as artificial stone) in San Francisco. He first used reinforcing in 1877, and in 1884 he patented a system using twisted square rods to help the development of bond between the concrete and reinforcing.[8] His largest work of the time was the Leland Stanford, Jr. Museum at Stanford University, the first building to use exposed aggregate. He was also responsible for several industrial buildings in New Jersey and. Pennsylvania, such as the 1903-1904 construction of the Kelly and Jones Machine Shop (Figure 1.3) in Greensburg, Pennsylvania.

FIGURE 1.3   Kelly and Jones Machine Shop. (Portland Cement Association)

The Ingalls Building (Figure 1.4), a landmark structure in Cincinnati, was built in 1904 using a variation of the Ransome system. Designed by the firm of Elzner and Henderson, it was the first concrete skyscraper, reaching 16 stories and 210 ft.

On the other side of the Atlantic, François Hennebique, a successful mason turned contractor in Paris, had started to build reinforced concrete houses in the late 1870s. He took out patents in France and Belgium for the Hennebique system of construction and proceeded to establish an empire of franchises in major cities. He promoted the material by holding conferences and developing standards within his own company network. Most of his buildings (like Ransome's) were industrial.

When the far-flung company was at its peak, Hennebique was fulfilling more than 1500 contracts annually.[9] More than any other individual, he was responsible for the rapid growth of reinforced concrete construction in Europe.

1.4   A REINFORCED CONCRETE ARCHITECTURE

If Hennebique was responsible for the acceptability of reinforced concrete as a building material, then it was Auguste Perret

FIGURE 1.4   The Ingalls Building (Elzner and Henderson). [*Portland Cement Association*]

FIGURE 1.5   Apartment building (Perret). [*Antione Spacagna*]

FIGURE 1.6   Theatre Champs Élysée (Perret). [*Guillaume Spacagna*]

FIGURE 1.7   Notre Dame du Raincy (Perret). [*Guillaume Spacagna*]

who made it acceptable as an architectural material. The works of Perret include not only factories and apartment buildings, but also museums, churches, and theaters. His better known works are in or around Paris, such as the delicately facaded apartment building at 25 bis Rue Franklin (Figure 1.5), completed in 1903. Just a few years later he designed the bulky, massive-looking, but spacious Theatre Champs Èlysée (Figure 1.6).

Notre Dame du Raincy (Figure 1.7), constructed in 1922, represented a significant departure from anything built in concrete before and is generally regarded as a masterpiece of architectural design. The lofty arched ceilings and the slender columns were very convincing statements as to the prowess of this newly accepted building material.

1.5   SHELL CONSTRUCTION

Reinforced concrete permitted the development of an entirely new building form, the thin shell. In 1930 Eduardo Torroja, the

FIGURE 1.8   Madrid Hippodrome (Torroja). [© Retoria/Y. Futagawa and Associated Photographers]

brilliant Spanish engineer, designed a low-rise dome of 3 ½ -in thickness and 150-ft span for the market at Algeciras, using steel cables for a tension ring. Torroja was also responsible for the statically elegant cantilevered stadium roof at the Madrid Hippodrome in 1935 (Figure 1.8).

At about the same time the Italian architect-engineer, Pier Luigi Nervi, began building his famous hangars for the Italian Air Force. At first these were cast in situ, but most of Nervi's work, including the Exhibition Hall at Turin (Figure 1.9) and the two sports palaces in Rome, was primarily of precast construction.

The master of the concrete shell, without dispute, would be the Spanish-born mathematician-engineer-architect, Felix Candela. Practicing mostly in Mexico City, he designed the Cosmic Ray Laboratory, with a 5/8-in-thick shell roof, for the University of Mexico City. He adopted the hyperbolic paraboloid form as his trademark and, making use of favorable labor costs, built many factories and churches in and around Mexico City using this form. His most striking building is the restaurant at Xochimilco (Figure 1.10), built in 1958, consisting of six identical paraboloid vaults.

FIGURE 1.9 Turin Exhibition Hall (Nervi). [*Hoepli S.p.A. Milan*]

1.6 FURTHER USES OF CONCRETE IN MODERN ARCHITECTURE

As a young architect Le Corbusier worked part-time in Perret's office but was always at odds with his employer, having no use for the espoused classical basis for design.[10] Le Corbusier was later to become the most highly regarded architect of the modern era, building almost exclusively in reinforced concrete. Among his celebrated works are the Villa Savoye (of flat plate construction, 1931) (Figure 1.11), the housing blocks on *pilotis* at Nantes and Marseille (late 1940s), the Chapel at Ronchamp (with walls of concreted masonry construction, 1957), the monastery of La Tourette (1959) (Figure 1.12), and the government complex at Chandigarh in India (1961). More so than his contemporaries, Le Corbusier was involved with the play of natural light as a design element, and concrete with its variable surface texture provided an excellent medium for his efforts.

FIGURE 1.10
Restaurant at
Xochimilco (Candela).
[*Agustin Estrada*]

Frank Lloyd Wright declared the prime assets of reinforced concrete to be its formability and monolithic property of construction, but he did not take advantage of this until late in his career. He was the first to exploit the cantilever as a design feature made possible by the continuous nature of reinforced concrete construction. The Kaufman House (Fallingwater) (Figure 1.13), built in 1936, is a tour de force in the use of the cantilever. Thin slabs seem to project beyond the possible, perhaps constructed containing as much steel as concrete!

In 1919 Mies van der Rohe had proposed the idea of a structural core for a high-rise building with cantilevered floor slabs,[11] but it

FIGURE 1.11   Villa Savoye (Le Corbusier). [*Guillaume Spacagna*]

FIGURE 1.12   La Tourette (Le Corbusier). [*Bernard Fleury*]

FIGURE 1.13   The Kaufman House (Wright). [*Harold Corsini, The Western Pa. Conservancy*]

FIGURE 1.14 Johnson Wax Tower (Wright). [*S. C. Johnson & Sons, Inc.*]

was not until 1947 that Wright brought the idea to fruition with his design for the Johnson Wax Tower (Figure 1.14) at Racine, Wisconsin. The entire Johnson Wax headquarters complex was hailed as being among the best of Wright's creations.

Wright's claim to an organic basis for his designs and the need to exploit the "plastic" nature of reinforced concrete reached a high point with his design of the Guggenheim Museum in 1956. The monumental spiral form became an overnight New York City landmark.

1.7   HIGH-STRENGTH CONCRETE AND HIGH-RISE BUILDINGS

High-rise construction in concrete progressed slowly forward from the Ingalls Building in

FIGURE 1.15 Marina City (Goldberg). [*Portland Cement Association*]

1904. The giants and midgiants of the 1930s were all of steel construction. The Johnson Wax Tower, however, provided the impetus for Bertrand Goldberg's twin towers of Marina City (Figure 1.15), though on a vastly different scale. The Chicago 60-story high-rise, erected in 1962, heralded the beginning of the use of reinforced concrete in modern skyscrapers and with it, competition for the steel frame. Place Victoria in Montreal, constructed in 1964, reached a height of 624 ft utilizing

FIGURE 1.16   One Shell Plaza (SOM). [*Portland Cement Association*]

FIGURE 1.17   Scotia Plaza Building (Webb, Zerafa, Menkes Housden). [*Campeau Corporation*]

6000 psi concrete in the columns. Concretes of higher strength proved to be the key to increased height, permitting as they do a reasonable column size on the floors below. One Shell Plaza (Figure 1.16) in Houston topped out at 714 ft in 1970 using 6000 psi concrete. The Chicago area, with its plentiful supply of high-quality fly ash (which helps to achieve a more workable concrete at lower water/cement ratios), has spawned the greatest concentration of tall reinforced concrete buildings. The 70-story Lake Point Towers used 7500 psi concrete to reach 645 ft in 1968. Water Tower Place reached 859 feet in 1973 with concrete strengths as high as 9000 psi thanks to a superplasticizing admixture (see Section 2.5).

In 1989 the Scotia Plaza Building (Figure 1.17) in Toronto was completed to a height of 907 ft. In 1990 two more towers in Chicago exceeded 900 ft. The taller of these is the building at 311 S. Wacker Drive shown next to the Sears Tower in Figure 1.18.

During the 1990s, the high-rise building field was dominated by Asian countries. In 1992, the Central Tower office building was constructed in Hong Kong to a height of 1150 feet. The Citic Plaza office tower was constructed in Guangzhou, China in 1997. It is currently the world's tallest reinforced concrete building with a height of 1210 feet. Both of these structures were designed by the Hong Kong firm of Dennis Lau and Ng Chunman, Architects and Engineers. The Petronas twin office towers in Kuala Lampur, Malaysia designed by Cesar Pelli Associates topped out at 1400 feet in 1998. It was constructed using

FIGURE 1.18   311 S. Wacker Drive (Kohn/ Pederson/Fox). [Joe Aker, Houston, Texas]

FIGURE 1.19   Taipei 101 (C.Y. Lin & Partners) [emporis.com]

a hybrid approach in which large steel tubes are filled with and surrounded by reinforced concrete. Also in 1998, the 1300-foot Jin Mao Tower in Shanghai was constructed using a similar technique. It is a mixed-use building designed by SOM with office, retail and hotel space. The concrete used in the lower story columns of such structures generally has a compressive strength in the 10 000 to 14 000 psi range.

While steel used to be the dominant construction material for tall buildings, it is now reinforced concrete, used either compositely with steel or with regular reinforcing. This is because in really tall structures, the added stiffness afforded by larger columns is necessary to limit sway due to wind. This is particularly important in residential buildings because humans are less tolerant of motion over longer periods of time. An added bonus with reinforced concrete is the additional safety achieved in terms of fire and security. This is not to imply that such buildings could survive an airplane impact, but they would be better than steel buildings at sustaining attacks by missiles or car bombs.

At this writing, the tallest building in the world is Taipei 101 (101 floors) in Taiwan. It is 1570 feet tall and also uses hybrid construction with numerous columns made of steel tubes approximately four-feet square. These have reinforced concrete both inside and out. Taipei 101 is a mixed-use building but includes no residential floors. It has a pagoda

FIGURE 1.20
Burj Dubai (SOM)
[wikipedia.org,
EMAAR Properties]

theme as seen in Figure 1.19. The C.Y. Lin & Partners design also reflects the large baskets used in the region to carry products to and from produce markets. It is appropriate to say "at this writing," because taller buildings are being designed or constructed at this time. Funds are being pursued to construct the Fordham Spire in Chicago, designed by Santiago Calatrava. It is to be a residential reinforced concrete building, planned for a height of approximately 2000 feet, and is scheduled for completion in 2009. Currently under construction is the Burj Dubai building in Dubai, United Arab Emirates. (Burj means tower.) A rendering is shown in Figure 1.20. It was designed by SOM and is a mixed-use reinforced concrete building with offices, apartments and a hotel. It will be completed in 2008 having 160 floors and an approximate height of 2300 feet!

For up-to-date information on tall buildings, one of the better websites is www.skyscraperpage.com.

REFERENCES

1. Axel Böethius and J. B. Ward-Perkins, *Etruscan and Roman Architecture*, Penguin Books, Middlesex, England, 1970, pp. 246-247.

2. Rowland J. Mainstone, *Developments in Structural Form*, The MIT Press, Cambridge, MA, 1975, p. 116.

3. Peter Collins, Concrete, *The Vision of a New Architecture*, Faber and Faber, London, 1959, pp. 19-24.

4. W. Fisher Cassie, "The First Structural Reinforced Concrete," Structural Concrete, Vol. 2, No. 10, July-August, 1965, p. 382.

5. Carl W. Condit, *American Building, Materials and Techniques from the First Colonial Settlements to the Present*, University of Chicago Press, Chicago, 1968. p. 160.

6. J. E. C. Farebrother, "Concrete-Past, Present, and Future," *The Structural Engineer*, October 1962, p. 339.

7. Collins, p. 61.

8. Condit, p. 171.

9. Collins, p. 72.

10. Collins, p. 153.

11. Arthur Drexler, *Ludwig Mies van der Rohe*, George Braziller, New York, 1960, p. 14.

2

CHARACTERIS-TICS
OF CONCRETE
AND
REINFORCING

2.1 PROPERTIES OF CONCRETE

Concrete is an artificial conglomerate stone made essentially of Portland cement, water, and aggregates. When first mixed the water and cement constitute a paste which surrounds all the individual pieces of aggregate to make a plastic mixture. A chemical reaction called *hydration* takes place between the water and cement, and concrete normally changes from a plastic to a solid state in about 2 hours. Thereafter the concrete continues to gain strength as it *cures*. A typical strength-gain curve is shown in Figure 2.1. The industry has adopted the 28-day strength as a reference point, and specifications often refer to compression tests of cylinders of concrete which are crushed 28 days after they are made. The resulting strength is given the designation f'_c.

During the first week to 10 days of curing it is important that the concrete not be permitted to freeze or dry out because either of these occurrences would be very detrimental to the strength development of the concrete. Theoretically, if kept in a moist environment, concrete will gain strength forever; however, in practical terms, about 90% of its strength is gained in the first 28 days.

Concrete has almost no tensile strength (usually measured to be about 10 to 15% of its compressive strength), and for this reason it is

FIGURE 2.1 Typical strength-gain curve.

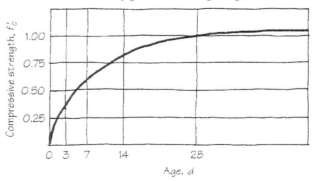

almost never used without some form of reinforcing. Its compressive strength depends upon many factors, including the quality and proportions of the ingredients and the curing environment. The single most important indicator of strength is the ratio of the water used compared to the amount of cement. Basically, the lower this ratio is, the higher the final concrete strength will be. (This concept was developed by Duff Abrams of The Portland Cement Association in the early 1920s and is in worldwide use today.) A minimum *w/c ratio* (water-to-cement ratio) of about 0.3 by weight is necessary to ensure that the water comes into contact with all cement particles (thus assuring complete hydration). In practical terms, typical values are in the 0.4 to 0.6 range in order to achieve a workable consistency so that fresh concrete can be placed in the forms and around closely spaced reinforcing bars.

Typical stress-strain curves for various concrete strengths are shown in Figure 2.2. Most structural concretes have f_c' values in the 3000 to 5000 psi range. However, lower-story columns of high-rise buildings will sometimes utilize concretes of 12 000 or 15 000 psi to reduce the column dimensions which would otherwise be inordinately large. Even though Figure 2.2 indicates that the maximum strain that concrete can sustain before it crushes varies inversely with strength, a value of 0.003 is usually taken (as a simplifying measure) for use in the development of design equations.

Because concrete has no linear portion to its stress-strain curve, it is difficult to measure a proper modulus of elasticity value. For concretes up to about 6000 psi it can be approximated as

$$E = 33w^{1.5}\sqrt{f_c'} \qquad (2\text{-}1)$$

where w = unit weight (pcf)

f_c' = cylinder strength (psi)

(It is important that the units of f_c' be ex-

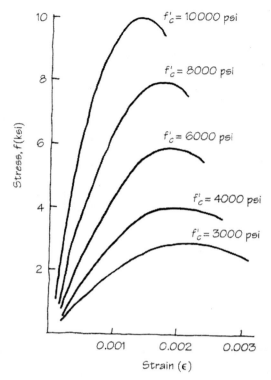

FIGURE 2.2   Stress versus strain curves.

pressed in psi and not ksi whenever the square root is taken.) The weight density of reinforced concrete using normal sand and stone aggregates is about 150 pcf. If 5 pcf of this is allowed for the steel and w is taken as 145 in Equation (2-1), then

$$E = 57\,000\sqrt{f_c'} \qquad (2\text{-}2)$$

E values thus computed have proven to be acceptable for use in deflection calculations.

As concrete cures it shrinks because the water not used for hydration gradually evaporates from the hardened mix. For large continuous elements such shrinkage can result in the development of excess tensile stress, particularly if a high water content brings about a large shrinkage. Concrete, like all materials, also undergoes volume changes due to thermal effects, and in hot weather the heat from the exothermic hydration process adds to this problem. Since concrete is weak in tension, it will often develop cracks due to such shrinkage and temperature changes. For example, when a freshly placed concrete slab-on-grade expands due to temperature change, it develops internal compressive stresses as it overcomes the friction between it and the ground surface. Later when the concrete cools (and shrinks as it hardens) and tries to contract, it is not strong enough in tension to resist the same frictional forces. For this reason contraction joints are often used to control the location of cracks that inevitably occur and so-called temperature and shrinkage reinforcement is placed in directions where reinforcing has not already been specified for other reasons. The purpose of this reinforcing is to accommodate the resulting tensile stresses and to minimize the width of cracks that do develop.

In addition to strains caused by shrinkage and thermal effects, concrete also deforms due to creep. *Creep* is increasing deformation that takes place when a material sustains a high stress level over a long time period. Whenever constantly applied loads (such as dead loads) cause significant compressive stresses to occur, creep will result. In a beam, for example, the additional long-term deflection due to creep can be as much as two times the initial elastic deflection. The way to avoid this increased deformation is to keep the stresses due to sustained loads at a low level. This is usually done by adding compression steel. Creep deflection is examined in more detail in Chapter 9.

2.2 MIX PROPORTIONS

The ingredients of concrete can be proportioned by weight or volume. The goal is to provide the desired strength and workability at minimum expense. Sometimes there are special requirements such as abrasion resistance, durability in harsh climates, or water impermeability, but these properties are usually related to strength. Sometimes concretes of higher strength are specified even though a lower f_c' value would have met all structural requirements.

As mentioned previously, a low water-to-cement ratio is needed to achieve strong concrete. It would seem therefore that by merely keeping the cement content high one could use enough water for good workability and still have a low w/c ratio. The problem is that cement is the most costly of the basic ingredients. The dilemma is easily seen in the schematic graphs of Figure 2.3.

Since larger aggregate sizes have relatively smaller surface areas (for the cement paste to coat) and since less water means less cement, it is often said that one should use the largest practical aggregate size and the stiffest practical mix. (Most building elements are constructed with a maximum aggregate size of ¾ to 1½ in, larger sizes being prohibited by the closeness of the reinforcing bars.)

A good indication of the water content of a mix (and thus the workability) can be had from a standard *slump test*. In this test a metal

FIGURE 2.3 Mix proportion relationships.

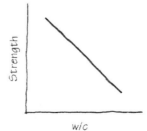 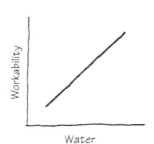 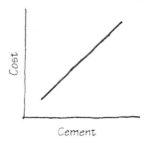

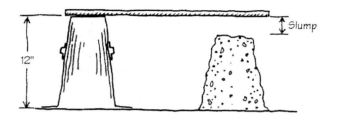

FIGURE 2.4 Slump test.

cone 12 in tall is filled with fresh concrete in a specified manner. When the cone is lifted, the mass of concrete "slumps" downward (Figure 2.4) and the vertical drop is referred to as the slump. Most concrete mixes have slumps in the 2- to 5-in range.

2.3 PORTLAND CEMENT

The raw ingredients of Portland cement are iron ore, lime, alumina, and silica, which are used in various proportions depending upon the type of cement being made. These are ground up and fired in a kiln to produce a *clinker.* After cooling, the clinker is very finely ground (to about the texture of talcum powder) and a small amount of gypsum is added to retard the initial setting time. There are five basic types of Portland cement in use today:

Type I   General purpose
Type II   Sulfate resisting, for use where the concrete will be in contact with high sulfate soils
Type III   High early strength, which gains strength faster than Type I, enabling forms to be removed sooner
Type IV   Low heat of hydration, for use in massive construction
Type V   Severe sulfate resisting

Type I is the least expensive and is used for the majority of concrete structures. Type III is also frequently employed because it enables forms to be reused quickly, allowing construc-

tion time to be reduced. (It is important to note that while Type III gains strength faster than Type I, it does *not* take its initial set any sooner.) Type IV has not been produced for a number of years because the low heat of hydration can be achieved with the addition of fly ash or slag to the mix.)

2.4 AGGREGATES

Fine aggregate (sand) is made up of particles which can pass through a ⅜-in sieve; coarse aggregates are larger than ⅜ in in size. Aggregates should be clean, hard, and well-graded, without natural cleavage planes such as those that occur in slate or shale. The quality of aggregates is very important since they make up about 60 to 75% of the volume of the concrete; it is impossible to make good concrete with poor aggregates. The grading of both fine and coarse aggregate is very significant because having a full range of sizes reduces the amount of cement paste needed. Well-graded aggregates tend to make the mix more workable as well.

Normal concrete is made using sand and stones, but lightweight concrete can be made using industrial by-products such as expanded slag or clay as lightweight aggregates. This concrete weighs only 90 to 125 pcf and high strengths are more difficult to achieve because of the weaker aggregates. However, considerable savings can be realized in terms of the building self-weight, which may be very important when building on certain types of soil. Insulating concrete is made using perlite and vermiculite; it weighs only about 15 to 40 pcf and has no structural value.

2.5 ADMIXTURES

Admixtures are chemicals which are added to the mix to achieve special purposes or to meet certain construction conditions. There are basically four types: air-entraining agents, workability agents, retarding agents, and accelerating agents.

In climates where the concrete will be exposed to freeze-thaw cycles air is deliberately mixed in with the concrete in the form of billions of tiny air bubbles about 0.004 in in diameter. The bubbles provide interconnected pathways so that water near the surface can escape as it expands due to freezing temperatures. Without air-entraining, the surface of concrete will almost always spall off when subjected to repeated freezing and thawing. (Air-entraining also has the very beneficial side effect of increasing workability without an increase in the water content.) Entrained air is not to be confused with entrapped air, which creates much larger voids and is caused by improper placement and consolidation of the concrete. Entrapped air, unlike entrained air, is never beneficial.

Workability agents, which include water-reducing agents and plasticizers, serve to reduce the tendency of cement particles to bind together in flocs and thus escape complete hydration. Fly ash, a by-product of the burning of coal that has some cementitious properties, is often used to accomplish a similar purpose. Sometimes slag and silica are added along with the fly ash. To obtain very high strength concrete, superplasticizers are frequently used. These are chemicals which when added to a mixture serve to increase the slump greatly, making the mixture very soupy for a short time and enabling a low-water-content (or otherwise very stiff) concrete to be easily placed. Superplasticizers are responsible for the recent development of very high strength concretes, some in excess of 15 000 psi because they greatly reduce the need for excess water for workability.

Retarders are used to slow the set of concrete when large masses must be placed and the concrete must remain plastic for a long period of time to prevent the formation of "cold joints" between one batch of concrete and the next batch. Accelerators serve to increase the rate of strength gain and to decrease the initial setting time. This can be beneficial when concrete must be placed on a steep slope with a single form or when it is desirable to reduce the time period in which concrete must be protected from freezing. The best known accelerator is calcium chloride, which acts to increase the heat of hydration, thereby causing the concrete to set up faster.

Other types of chemical additives are available for a wide range of purposes. Some of these can have deleterious side effects on strength gain, shrinkage, and other characteristics of concrete, and test batches are advisable if there is any doubt concerning the use of a particular admixture.

2.6 THE ACI CODE

The American Concrete Institute (ACI), based in Detroit, Michigan, is an organization of design professionals, researchers, producers, and constructors. One of its functions is to promote the safe and efficient design and construction of concrete structures. The ACI has numerous publications to assist designers and builders; the most important one in terms of building structures is entitled *Building Code Requirements for Reinforced Concrete and Commentary*. It is produced by Committee 318 of the American Concrete Institute and contains the basic guidelines for building code officials, architects, engineers, and builders regarding the use of reinforced concrete for building structures. Information is presented concerning materials and construction practices, standard tests, analysis and design, and structural systems. This document has been adopted by most building code authorities in the United States as a standard reference. It provides all rules regarding reinforcing sizes, fabrication, and placement and is an invaluable resource for both the designer and the detailer.

Over the years, periodic updates have been made to this document and this text makes constant reference to the 2005 edition. Documents and officials refer to it by its numerical designation ACI 318-05.

2.7 TYPES OF REINFORCING

The most commonly used type of reinforcing is deformed steel bars. These are circular rods, often called *rebars*, which come in a range of nominal diameters from ⅜ to 1⅜ in; two larger sizes are also produced that have nominal diameters of 1¾ and 2¼ in. Bar sizes are designated by numbers which represent the number of eighths of an inch in the nominal diameter; thus, a #9 bar has a nominal diameter of 1⅛ in. Bar sizes are given as #3 through #11, then #14 and #18 for the special bars. For metric values, the approximate diameter in millimeters is used. In that case, bar sizes are #10 through #36 plus #43 and #57 for the two larger sizes. The precise cross-sectional areas cannot be determined using these numbers because of the deformation patterns. Table A.1 of the Appendix gives the bar sizes in both inch-pound units and their metric equivalents and the cross-sectional areas in the inch-pound system.

Three strengths of steel are used for making rebars used in building construction today, Grades 40, 60, and 75. The numbers refer to the yield strength (ksi) in each case. While Grade 40 bar was the industry standard for many years, today it is almost impossible to get that grade in anything but the smaller bar sizes such as #3 and #4. Grade 40 steel is easier to bend than Grade 60, and since these smaller bars can be bent in the field without elaborate equipment, there is reason to continue manufacturing these in the lower-grade steel. Grade 75 is used infrequently and then mostly in lower story columns. The examples and problems in this book use Grades **40** and **60**.

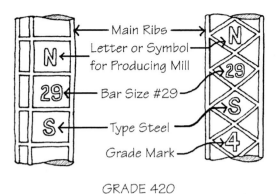

GRADE 420

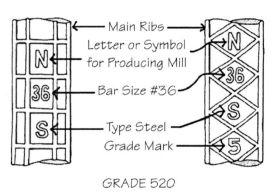

GRADE 520

FIGURE 2.5 Typical reinforcing bar markings.

Figure 2.5 shows some typical rebar markings required of U.S. producers. The first mark is a letter or symbol representing the manufacturing company. The second mark indicates the bar size using metric values such as 25 and 43. The third mark represents the type of steel used and is most commonly the letter S which indicates new billet steel. The fourth mark indicates the strength of the steel, i.e., Grade 40, 60, or 75. The industry refers to these values by their metric equivalents, i.e., 280, 420, and 520 Mpa, respectively. Grade 40 reinforcing steel does not bear a mark indicating the strength, whereas 420 Mpa ($f_y = 60$ ksi) steel has the mark 4 and 520 Mpa ($f_y = 75$ ksi) has the mark 5. Manufacturers are allowed an alternate way of indicating the two higher strengths. All bars have two longitudinal ribs, one on each side of the bar. Instead of the number 4, they may use *one* additional longitudinal rib instead of the number 5, they may use *two* additional ribs.

Other materials can be used to make reinforcing bars, and over the years builders have worked at various times with bamboo and other natural fibers. The choice of material is limited, however, because the alkaline environment created by the cement will cause deterioration of some materials (such as fiberglass) and most metals do not have a thermal coefficient of expansion compatible with that of concrete. The result is that steel is used by all industrialized nations today.

For slabs, especially slabs-on-grade, a special type of reinforcing called *welded wire fabric* can be used. It consists of a coarse mesh of wires usually 4 or 6 in on center. The old designation would call out the spacing followed by the wire gauge size, e.g., 6 × 6-10/10 would mean a 6-in-square grid using 10-gauge wire. The preferred designation gives the cross-sectional area of the wire in hundredths of a square inch, so the preceding designation is now given as 6 × 6-W1.4 × 1.4. If the wire is deformed instead of smooth, the W is replaced by a D.

Since about 1970 experiments have been conducted using short fibers dispersed throughout the concrete mix instead of reinforcing bars. Steel has received the most attention in the form of short lengths of wire ½ to 1½ in long and 0.004 to 0.1 in in diameter. Most versions are deformed along the length or crimpled at the ends to improve bond. Tests have shown that the use of such fibers increases the tensile strength relative to that of plain concrete (as one might expect) and markedly improves both ductility and impact resistance. Shrinkage characteristics are also greatly improved.

Researchers have concerned themselves with overcoming two major problems– assuring proper dispersion and maintaining sufficient workability. Both. issues are related to the *L*-over-*d* ratio of the individual fibers, which must be kept low. Workability also decreases as the percentage of fibers increases and it is difficult to make beams and columns with strength capacities as great as those utilizing regular reinforcing bars.

In some parts of the United States polypropylene fibers (about the same length as the steel ones but much thinner) have been found to be more effective than welded wire fabric for shrinkage crack control in slabs-on-grade.

2.8 DEVELOPMENT OF BOND

Regardless of the type of reinforcing used, the proper development of bond between the reinforcing and concrete is essential to achieve the necessary load capacity of reinforced concrete structures. The basic premise of flexural behavior in reinforced concrete is that the tensile strain of the two materials be identical so that the steel can develop the stress needed to develop the tensile force resultant of the internal moment couple. (This is explained further in Chapter 6.) It is impossible for the strains to be the same unless there is an adequate bond between the two materials. Figure 2.6 shows what would happen in a beam if the bond did not adequately develop. The reinforcing bars in the lower part of the beam would not be strained and would not elongate as the bottom fibers of the concrete went into tension. Under very low load the tensile concrete would, of course, crack and the beam would collapse as the steel slipped inside the concrete, restrained only by surface friction. The development of bond occurs from three sources:

1. Chemical adhesion between the steel and concrete

2. Mechanical forces (static friction)

3. Surface deformations of the bars

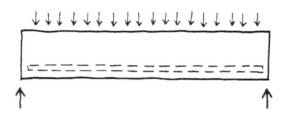

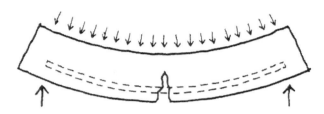

FIGURE 2.6   Failure resulting from a lack of bond.

Modern deformed bars provide enough bond so that in many cases hooks or bends at the ends of the bars, required in the past, are no longer needed. The ends of bars are bent into L-shaped or U-shaped hooks only when there is insufficient distance at the end of a beam or slab to embed the steel.

Determining the magnitude of the stresses between tensile reinforcing bars and the adjacent concrete is difficult because the situation is complicated by the presence of numerous cracks in the concrete. Researchers have proven that bond stress varies markedly in the immediate vicinity of such cracks and the ultimate failure mechanism is a complex one involving a "wedging" action of the concrete in compression and shear due to the deformations on the bars. The ACI Code uses the concept of *development length*, which is defined as the embedment length required of a straight bar to develop its full yield strength. In other words, the bar, if embedded at least that far into the concrete, will yield in tension before pulling out. There is also a development length for compression but it is usually not as large and is easier to meet in terms of

detailing the proper bar lengths.

The development length is a function of bar size and location. Naturally larger bars require longer embedment lengths because of their larger yield forces. Bars near the top of concrete elements, such as the tensile reinforcement at the ends of continuous beams, require greater development lengths because it has been found that weak concrete sometimes occurs at the underside of such bars as entrapped air and excess water "bleed" to the surface. It can also be noted that because bond failure is related to the concrete strength, development length decreases (at least for the larger bars) as concrete strength goes up.

2.9 REINFORCING PLACEMENT

Because the moment varies over the length of a beam, the area of steel needed to carry the tensile force resultant can also vary. For example, in a uniformly loaded simple beam, there is no reason to carry the full amount of steel needed for the midspan maximum moment all the way to the ends of the beam where the moment is zero. As the moment decreases some of the bars can be deleted or "cut off" provided a symmetrical pattern is maintained with respect to the cross section. For positive moment steel the Code stipulates that a certain fraction of the steel be carried all the way to the end of beam; for simply supported beams this fraction is ⅓, and for continuous members it is ¼. Both positive and negative steel have theoretical *cutoff* points where a portion of the steel can be dropped off. A typical requirement calls for such steel to be continued a distance equal to 12 bar diameters or the effective beam depth, whichever is larger, beyond where the moment value indicates it would no longer be needed. Figure 2.7 provides a qualitative look at the changing amounts of reinforcing over the span as the moment varies. For the top and bottom steel, two layers are shown at each location for

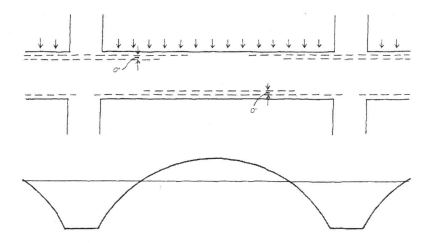

FIGURE 2.7 Typical reinforcing bars for moment in a continuous beam.

clarity, but the zero dimension indicates that the steel bars in both the top and bottom actually exist in a single layer. With deeper beams, sometimes the steel must be placed in multiple layers to accommodate the beam width.

There are several other Code requirements involving the location of cutoff points, as it is recognized that the stress in the remaining bars must suddenly increase and that the termination of a bar promotes the formation of shear cracks. There are also specific rules governing the splicing of bars in both tension and compression. Since this is not a text written as a treatise on the detailing of reinforced concrete, the reader is referred directly to the ACI Code for further information. The determination of the depth and amount of steel needed at any section for moment, however, is treated thoroughly in Chapters 6 and 7.

In spandrel (perimeter) beams, the ACI Code requires that a portion of the negative steel be carried over the full beam length as is required for the positive steel in all beams. This assures the some of both the top and bottom steel will completely "wrap" the building perimeter.

In addition to the moment steel, reinforcing is also provided to combat shearing forces.

This is usually in the form of stirrups of #3 or #4 rebars placed vertically around the outside of the moment steel. Figure 2.8 shows how the spacing between stirrups increases as the shear forces become smaller. Frequently the two corner bars of top steel (such as that shown in Figure 2.7) will be carried all the way across the center of the beam to help locate the stirrups if any are required in this region and to create a more rigid "cage" of reinforcing steel for ease of construction.

All reinforcing must be protected from the effects of moisture and fire by concrete cover. The typical requirement for the clear cover distance between *any* steel reinforcement and the surface of the concrete is $1\frac{1}{2}$ in for beams and columns and $\frac{3}{4}$ in for slabs. This minimum clear cover goes to 2 in (for #6 bars and larger) if the surface is exposed to the weather or if earth will later be placed against the concrete and 3 in for all steel if the concrete is actually cast against the earth.

In addition, the Code recognizes that minimum spacing requirements are needed so that bars will not be detailed so close to one another as to inhibit the proper placement of the wet concrete, thereby causing voids and preventing proper bond. These are summarized for a beam in Figure 2.9. It goes almost without saying that bars in different rows in a

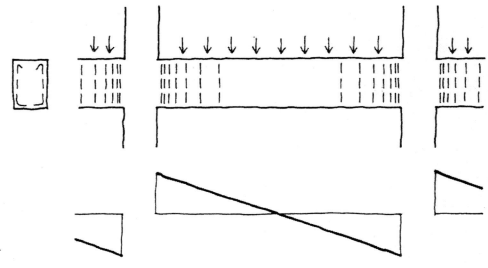

FIGURE 2.8 A typical stirrup pattern.

beam must be placed directly over and under each other; otherwise the placement of concrete would be almost impossible.

In larger columns and girders bars are sometimes bundled in groups of three or four by tying them together to act as one. In such cases the spacing requirements are different from those given earlier.

FIGURE 2.9 Clear distance requirements.

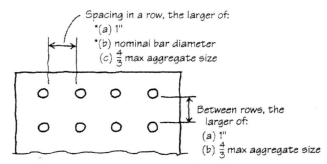

Spacing in a row, the larger of:
  *(a) 1"
  *(b) nominal bar diameter
  (c) $\frac{4}{3}$ max aggregate size

Between rows, the larger of:
  (a) 1"
  (b) $\frac{4}{3}$ max aggregate size

*For columns; $1\frac{1}{2}$" and $1\frac{1}{2}$ bar diameters, respectively.

3

REINFORCED CONCRETE BUILDING SYSTEMS

3.1 INTRODUCTION

Apart from fire resistance, one of the great advantages of reinforced concrete is its versatility in terms of construction geometry and form. No other building material provides the diversity of systems such as the examples shown in Figure 3.1. The inherent continuity that occurs when elements are cast monolithically contributes to the uniqueness of concrete and makes many of these forms possible. Tapered sections and curvilinear forms are constructed with ease relative to attempts in wood, steel, or masonry. This continuity also provides naturally occurring moment-resistant connections which enable members to cantilever without elaborate detailing. Cantilevers are not only attractive in terms of their visual boldness but can also result in greater efficiency. A cantilevered deck that projects out approximately 20 to 30% of the main span can actually result in smaller maximum design moments and a more open floor system.

The longer span shell and plate structures of Figure 3.1 are unique designs which usually require close collaboration in terms of architectural and engineering efforts right from the moment of conception. This text treats the behavior and preliminary design of the more traditional frame and deck systems which are described in some detail in this chapter. This is probably the easiest way for the novice designer to become familiar with bending moments, shears, and other forces that affect all concrete structures. Factors that influence the development of various types of stresses and deformations and the basis for providing adequate resistance to those effects are generally the same regardless of the complexity or uniqueness of the structure.

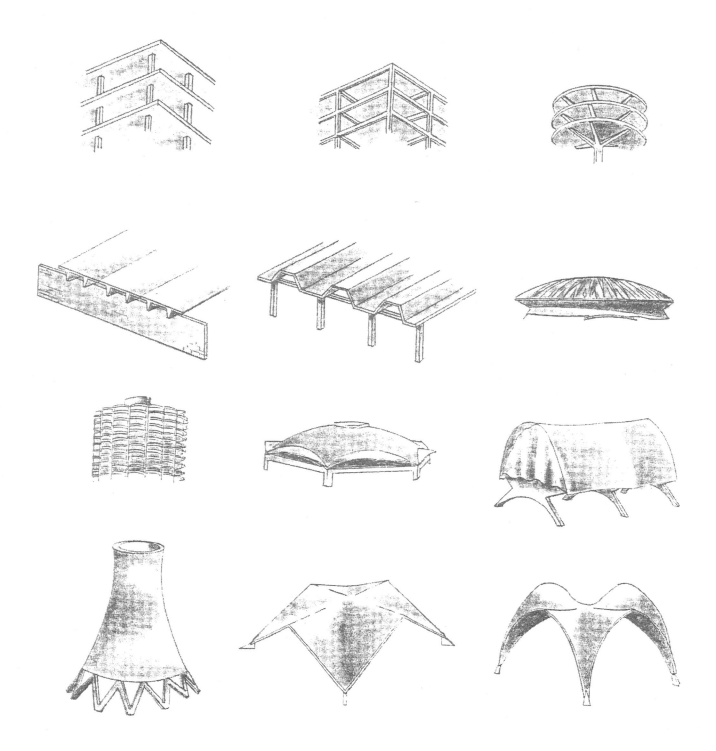

FIGURE 3.1 Examples of reinforced concrete building forms. (Credit: Nathan Goldberger)

3.2 FLAT PLATE

The simplest deck system is provided by the *two-way flat plate* illustrated in Figure 3.2. It has smooth parallel top and bottom surfaces without column capitals or drop panels and is often used with a square column grid. It is most effective for short spans with relatively light loads and is frequently used for motel and apartment construction. The smooth undersurface makes the two-way flat plate system very flexible in terms of partition placement and mechanical system layout.

It is called two-way because the reinforcing bars usually form an orthogonal grid of two layers of steel. Negative moment steel is placed near the top of the plate close to the columns and crossing the column lines (imaginary lines connecting the columns), and positive moment steel is used near the bottom surface out in the middle of the plate in both directions. Positive moment steel is also needed parallel to the column lines. For some applications it can be economically prestressed by using continuous posttensioned tendons embedded in the slab, undulating from near the bottom of the plate to near the top surface following the signs of the moments. (Prestressing is discussed in Chapter 15.)

Use of flat plates is limited by their shear

TABLE 3.1 *Flat Plate Thicknesses for Residential Loads*

Span (ft)	Thickness (in)
12	5
15	6
18	7
20	8

capacity around the columns. *Punching shear* in which the plate slides down the columns would be the most likely cause of failure in the event of overload. As heavier loads are imposed or spans increase the use of a *shearhead* around the columns or the addition of beams becomes necessary. It would then be properly called a *flat slab*. Table 3.1 provides some reasonable thickness estimates for various flat plate spans assuming residential loads. Spans above 20 or 22 ft are usually uneconomical.

3.3 FLAT SLAB WITH SHEAR HEADS

The *flat slab* systems shown in Figure 3.3 are very similar to flat plates. A shearhead in the form of a capital or drop panel has been added to counter the punching shear forces around the columns. The conical capitals are best used in open plan areas, where partitions do not have to fit around them and are usually used with circular columns. The drop panel is more common and usually about one-third the slab span in size and several inches thicker than the main slab. Like the flat plate, this system is most effective with square or nearly square column bays. It is suitable for the heavier loads encountered in office or industrial buildings.

FIGURE 3.2 Flat plate

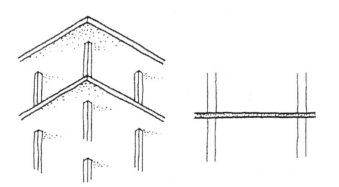

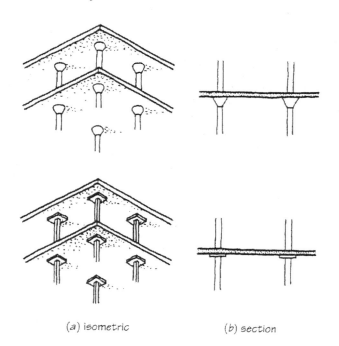

(a) isometric   (b) section

FIGURE 3.3 Flat slab with shearheads

The reinforcing pattern is very similar to that of a flat plate, and both systems have a significant advantage over the systems with beams in that it is feasible to locate a column several feet off the grid axis in plan (say, up to about 20% of the span). This can be a real asset in terms of planning flexibility as the spatial module is seldom a constant and columns can readily end up in awkward positions when they cannot be shifted. (It is obvious that when such an off-grid shift occurs, it must occur for all levels, not just for one.) Table 3.2 provides some approximate dimensions for drop panel slabs.

TABLE 3.2 *Flat Slab with Drop Panel Thicknesses*

Span (ft)	Thickness of Slab (in)	Total Depth of Beams (in)
16	6	8
20	8	11
24	10	14
28	12	17

3.4 FLAT SLABS WITH BEAMS

The two-way flat slab shown in Figure 3.4 can also be used with *beams* along the column lines instead of column shearheads. The slab can be slightly thinner but the overall thickness of the system is greater because the beams project downward farther than the shearheads. The system does not provide as much flexibility in terms of the mechanical layout because the ducts must run beneath the beams. However, the beams relieve the slab from the task of delivering the floor load to the columns; i.e., there is no punching shear. This means that it is possible to locate a vertical chase space adjacent to a column in the corner of the slab. This would be virtually impossible in the presence of capitals or drop panels.

FIGURE 3.4 Flat slab with beams

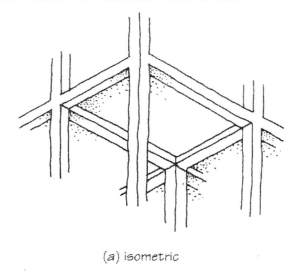

(a) isometric

The principal reason, however, for using a beam system is that the beams provide moment interaction with the columns; this interaction is essential if the frame is to resist lateral loads by rigid frame action. For most low-rise buildings these moment-resistant connections would be sufficient to withstand the lateral forces without having to use shear walls. The beams also may be needed to carry heavy partitions located above them. None of the thin plate or slab systems is very tolerant of point or heavy line loads.

Table 3.3 provides some reasonable dimensions for flat slab systems. Such systems become less efficient in the upper span ranges because of the self-weight penalty imposed by the slab thickness. Usually a waffle slab system is preferred for spans above about 24 ft. The amount of concrete needed is reduced significantly, which not only affects the design of the slab itself but also provides savings with the foundation, especially for taller structures.

It should be emphasized that these two-way systems are intended for use with more or less square column bays. If the bay becomes rectangular, say, with an aspect ratio of 3:2, and definitely if the ratio approaches 2:1, the slab behavior makes the transition from a two-way system to a one-way system. In such a case the slab will carry most of its load in the short direction to the nearest beam or column line and the overall thickness and the reinforcing in the short direction must increase.

TABLE 3.3 *Flat Slab with Beam Thicknesses*

Span (ft)	Thickness of Slab (in)	Total Depth of Beams (in)
20	7	16
24	9	20
28	11	25
30	13	28

3.5 WAFFLE SLAB

In the absence of prestressing the *waffle slab* is capable of achieving the largest spans of the conventional concrete deck systems. The two-way ribs are usually 2 to 3 ft on center and taper in width. Figure 3.5 shows two ways of transferring the shear to the columns, one using an effective shearhead and the other using beams. In either case the slab is merely made the full thickness in the appropriate places by deleting the metal pans that create the hollow coffers. Seldom is it necessary for the shearheads or beams to project below the slab. The system can be architecturally quite handsome and is often left exposed, although rarely entirely exposed because of the need for duct and pipe runs. Electrical and lighting requirements integrate well with the ribs and coffers but larger elements of the mechanical systems do not.

FIGURE 3.5 Waffle slab.

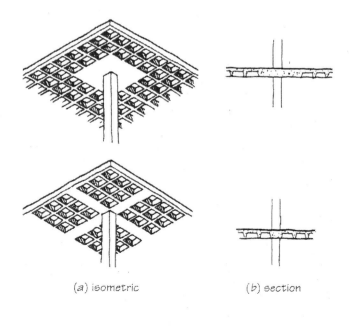

(a) isometric   (b) section

TABLE 3.4   *Waffle Slab Thicknesses*

Span (ft)	Thickness of Slab (in)	Total Depth of Beams (in)
25	3	12
30	3	15
35	4	19
40	5	24

The system can span up to 50 ft but is most often used in the 25- to 40-ft range. Because of the ribs it is relatively easy to create holes for mechanical shafts and skylights. Typical thicknesses are given in Table 3.4.

3.6 ONE-WAY SLAB

The *one-way* slab floor shown in Figure 3.6 is probably the most frequently used of any concrete system; it has great adaptability in terms of varying bay sizes and achieving an easy integration of spatial planning modules and structural planning modules. Beams can be placed on an irregular grid and the column spacing varied almost at will up to a maximum of about 36 ft. Above this spacing the girders become too deep and floor-to-floor height requirements become too great. The typical column spacing is about 26 to 30 ft.

Usually the slab itself forms a rectangle with the flexural steel running in the short direction; the only steel placed in the long direction is that needed for temperature and shrinkage. The typical slab span is 10 to 12 ft, with 14 to 16 ft being well within reason. Holes up to an entire bay in size can be located almost anywhere because it is merely necessary to place beams around the periphery of the hole to create stiff edges. The system can easily accommodate changing load patterns because the slab span can readily be reduced by the addition of intermediate beams.

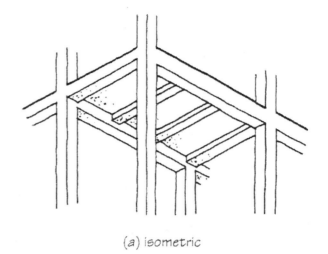

(a) isometric

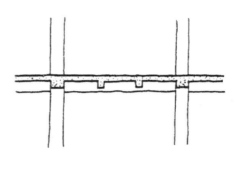

(b) section

FIGURE 3.6   One-way slab system.

The structural design of one-way slabs is treated in some detail in Chapter 11, where the American Concrete Institute (ACI) rules regarding thickness and reinforcing requirements are covered. After studying that chapter and the material on beam design in Chapter 7, the reader should be able to estimate reasonably the various depths required for preliminary design purposes.

As loads and spans increase the one-way pan joist system becomes a more efficient alternative to the one-way slab. Just as the two-way flat slab becomes less feasible for larger spans and is replaced by the waffle slab the pan joist system is used to reduce the amount of self-weight that a larger span one-way slab would have to accommodate.

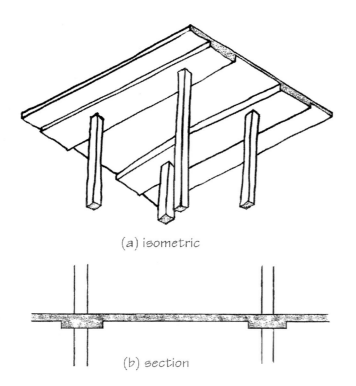

(a) isometric

(b) section

FIGURE 3.7 Banded beam system.

3.7 BANDED BEAM SYSTEM

Another one-way system called the banded beam is shown in Figure 3.7. It is actually less efficient structurally because (as is true with any beam of any material) beams should generally be deeper than they are wide. The banded beam system can be efficient in other ways, however. The overall depth will usually be less than that of a one-way slab system for similar spans. This means that ductwork and utilities can be more easily accommodated. It also means a reduction in the vertical distance from the ceiling on one story to the floor surface of the level above. This is of particular significance in high-rise buildings for two reasons: (1) because more floors can be constructed within a city's or district's height limitation, and (2) it saves energy by reducing

the amount of unoccupied heated and cooled space and the amount of exterior wall. The beams range in width from about 4 to 8 ft and about 8 to 16 inches in depth. Because of the span limitation on slabs discussed in the previous section, banded beam systems are more suited to rectangular column bays. The beams are usually prestressed over several spans; this increases the load capacity and reduces deflection, which can be a problem with any shallow system. Typical column spacings are 16 to 24 ft in the short direction and 24 to 34 ft in the direction of the beams.

3.8 PAN JOIST

The *pan joist* system shown in Figure 3.8 is a one-way system in concrete that echoes the prefabricated steel bar joist and wood joist systems. It is most often used for heavier load applications such as storage or industrial facilities. The ribs or joists are usually 18 to

FIGURE 3.8 Pan joist system.

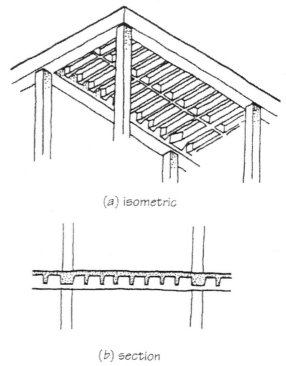

(a) isometric

(b) section

30 in on center and typical joist spans are in the 18- to 28-ft range. Naturally the system has a very strong visual directionality, and when used for public spaces, it is seldom left exposed because it would tend to influence the placement of partitions, lighting patterns, and other features of the space below. Like other joist systems, it employs a form of bridging usually at midspan of the joists. In this case the bridging is similar to a solid diaphragm running at right angles to the joists. This makes the pan space less usable for pipes and duct runs. As with any one-way joist system the designer has to be aware of the implications of running the joists the short direction or the long direction in a rectangular bay. If the joists run the short direction, they can be shallow but the beams will then usually be quite wide and deep. On the other hand, if the joists run the long way, they will be deeper, quite possibly about the same depth as the shorter but more heavily loaded beams. This has considerable implication in terms of the mechanical layout and partition placement. When the floor-to-floor height is not limited, most designers prefer to run the joists in the short direction. This means that the ducts and utilities will have to run under the deep long span beams, but the result is usually greater overall efficiency. Table 3.5 provides a rough idea of the depths required for such a system.

TABLE 3.5 *Pan Joist and Beam Depths*

Spans (ft)		Total Joist Depth (in)	Total Beam Depth (in)
Joist	Beam		
18	× 28	9	22
20	× 32	11	26
24	× 36	14	28
28	× 40	17	32

3.9 SYSTEM SELECTION

The selection of which system to use for different building types and different spaces is a complex issue. Novice designers always ask: How does one choose a structural system? These designers think that they should possess some sort of magical intuition or that there should be a set of rules or a "selection chart" of some sort. With little doubt, if certain architects do indeed have an intuitive sense of structural behavior and the appropriateness of different systems for different applications, they got this sense through *experience*. The young designer should expect to feel uncomfortable and at first must rely upon the experiences of others by *observing* and *asking*.

There is a tendency to ask: Which system will be the cheapest? But in this regard we caution that the cost of the above-grade structure of a building with normal spans is only about 15 to 20% of the total construction cost. This is usually less than the cost of the mechanical and electrical systems and about the same as the total cost of the interior finishes, ceiling system, and hardware. With long span systems this is certainly not the case and for buildings such as gymnasiums and concert halls the cost of the structural system is a dominant factor. For the great majority of buildings, however, the system should be selected for reasons other than cost.

It goes almost without saying that the function can influence the choice of structure; one certainly would not roof a bowling alley with a dome and it would probably be inappropriate to put a strongly defined and exposed one-way pan joist system above an expansive open space. The designer must be aware of the great influence that an exposed structure can have on the character of the space. It is *not* true that the structural module must be identical to the functional or planning module; there is a tendency on the part of inexperienced designers to think that columns should always be placed in corners or at the intersection of partitions, but this is seldom a realistic expectation. It *is* true, in general, that, small rectangular spaces call for some form of one-way system and that spaces which are more square are most compatible with two-way systems. Larger spans for department stores, office, and institutional buildings as well as structures supporting greater loads should probably use some sort of pan system, whether one- or two-way. Residential spaces with short spans and lighter loads can easily use simple flat plate systems. It is important to realize that the perceived height of a space is often heavily influenced by the character of its ceiling.

The designer should always be aware of the need for compatibility with the function. The structural system should lend order to forms and spaces and can serve well to modulate space that is otherwise lacking in organization, but it should not cause contortions and unreasonable modifications in terms of the use of the space. The designer should also be aware of two other building systems which must coordinate well with the structural system–the *circulation* system and the *mechanical* system. Clarity of people movement, both horizontally and vertically, within buildings is essential and the structure must not interfere with nor confuse this clarity. Since the mechanical and structural systems often need to occupy the same space, i.e., the space that lies between the finish ceiling and the finish floor, conflicts are sometimes inevitable. The skillful designer, however, can minimize these occurrences. It would seem that the mechanical system must be made to "fit around" the structure, but often an efficient and well-planned mechanical layout can establish the direction of joist and beam runs. In terms of overall building efficiency and life-cycle costs, the structural and mechanical systems should be coordinated at the preliminary stages of design.

3.10 LATERAL FORCE RESISTANCE

All tall buildings need to have special provisions for resisting wind or earthquake forces. With steel frames this is usually done through the use of diagonal bracing and/or the development of a tubular geometry. Diagonals are sometimes used with reinforced concrete frames as well but usually only when they also serve to provide the building exterior with a desired architectural image. Most often lateral resistance for high-rise concrete buildings is provided by some form of shear wall.

Figure 3.9 shows some examples of shear walls. Since solid walls are usually required for fire resistance purposes for vertical circu-

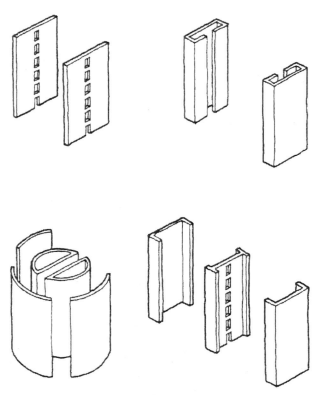

FIGURE 3.9 Examples of shear walls.

lation shafts, these shafts or cores are often incorporated into the lateral force resisting system. How many shear walls and how large they are is, for the most part, a function of geographical location (wind or seismic risk area), building height, and building shape. The design goal is to prevent a loss of stability by overturning or sliding and to prevent excess *drift* (lateral displacement).

Shear walls usually function as very deep, narrow cantilever beams sticking up out of the ground within the building envelope. Their design is often complicated by the presence of holes and irregular geometries necessitated by functional and circulation requirements.

Unlike timber and steel structures, lowrise concrete buildings, say, up to five or six stories, may not need to incorporate separate elements to handle lateral forces. The moment resistance provided by the continuity of the floor system-to-column connections may be sufficient to accommodate these loads. Again, this would depend upon the magnitude of the wind or earthquake forces and the geometry of the building.

Figure 3.10 shows a two-story frame resisting lateral forces by moment-resistant connections. The deformations have been greatly exaggerated for clarity.

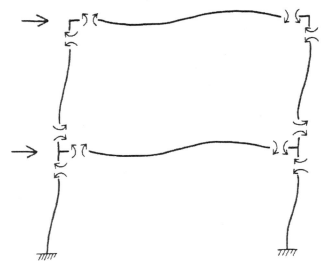

FIGURE 10 Moments at the connections resisting lateral forces.

4

APPROXIMATE FRAME ANALYSIS

4.1 INTRODUCTION

With the exception of precast construction, concrete structures are monolithic in nature, having continuity among slabs, beams, girders, and columns. Such structures are usually highly redundant and the proper determination of shears and moments requires some form of indeterminate analysis.

Continuous frames were analyzed by the *slope-deflection* method for 30 years or more starting before the turn of the century. This method involved the consideration of compatibility of the deformations among the various members meeting at each joint and required the laborious solution of many simultaneous equations. It was supplanted by the creation of the *moment distribution* method developed by Hardy Cross in the 1930s. This is an iterative technique based upon the relative stiffness of each of the frame members and is much less time-consuming than the slope-deflection method. It achieved universal popularity because of its simplicity and speed and remained the favored approach to indeterminate structures until the development of matrix algebra and high-speed computers some 30 years later. Today all but the simplest of continuous beams or frames are analyzed utilizing computer programs based upon matrix algebra.

Section 4.4 introduces the author's approach to estimating moments in continuous frames as opposed to determining them precisely. There is an often quoted adage about using computer software to do structural analysis and that is, "You have to know the answer before you can determine the answer." This sounds like nonsense but what is really being said here is that you have to have a rough idea of the expected outcome before performing a computer analysis. As addressed later, this enables you to detect gross errors in the machine-generated answers. (With today's highly sophisticated programs, such errors are almost always the result of poor input by the analyst.)

In terms of bending members, most design procedures require that the determination of the required cross-sectional dimensions be based upon the maximum *moment*. Stirrups are then placed as needed for shear and deflection is checked against code limitations. As with timber and steel beams, the moment is most often the controlling parameter.

During the schematic stage of building design when different spans and bay sizes are being considered, time constraints and the great number of variables involved make the performance of an accurate analysis infeasible. The ability of a designer to approximate rapidly critical design moments and consequent member sizes allows him or her to recognize and quickly discard irrational structural schemes, saving time and making the comparison of different framing schemes more meaningful. It also gives reasonable starting values of member sizes that can be used as input for the software to be used in the computer analysis. (All indeterminate techniques require the approximate stiffness of each frame member to be known; the stiffness of a member is directly related to its moment of inertia, which is, of course, dependent upon its section properties.) The designer who knows approximately what a moment or a beam depth should be is also more likely to spot errors that might occur during a formal analysis. Thus, the development of this kind of "educated intuition" can prove to be valuable at many stages during design; most importantly, perhaps it can stop an unsafe design from being carried to completion with probable disastrous consequences. Moment estimating helps the designer to develop this intuition.

4.2   LOADING PATTERNS

When several members share the task of supporting loads (as happens in every continuous structure), the resulting pattern of moments is based upon the relative stiffness (I/L) of the members. The stiffer the member is, the more moment it will take. For this reason it is difficult to establish general rules regarding the effect of load placement upon the development of maximum moments.

However, for structures of constant cross section (therefore, constant I), often the case for continuous beams over several supports, it is possible to make some predictions. In general, the *maximum negative moment* will develop if *adjacent spans* are loaded and the *maximum positive value* will occur when *alternate spans* are loaded. This is illustrated by the moment coefficients for a three-span beam, which are given in Figure 4.1. (These diagrams are plotted on the *tension side* of the reference line as is frequently done for continuous structures; i.e., if the bottom fibers of a beam are in tension, the resulting positive moment value will plot below the reference line rather than above it.) Notice that the absolute value of the maximum moment occurs in Figure 4.1*b* where one of the end spans has no load at all. The coefficient 0.117 indicates a moment approximately equal to $wL^2/8.5$. (It has to be realized, of course, that the dead load is always present, so this coefficient represents a theoretical upper limit rather than an actual maximum value.)

Using a similar diagram for a continuous beam with a larger number of spans, one can show that the effects of loads on nearby spans diminish rapidly with the distance from the span under consideration. For all practical purposes a large frame such as the

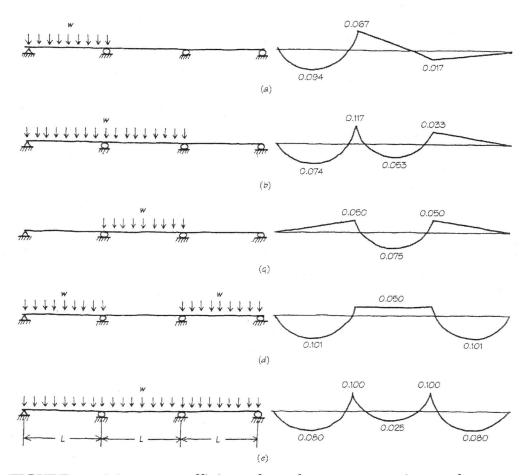

FIGURE 4.1 Moment coefficients for a three-span continuous beam. (M = coefficient $\times$ wL^2.)

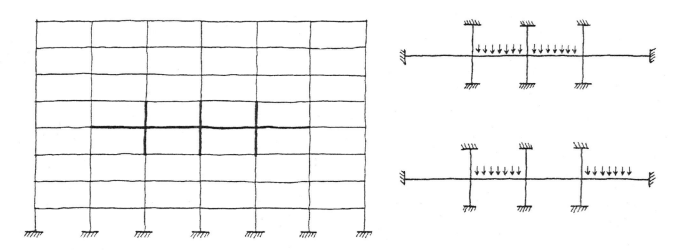

FIGURE 4.2 Patterns of live load to produce maximum design moments.

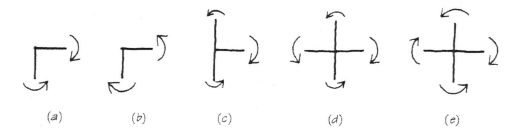

(a)   (b)   (c)   (d)   (e)

FIGURE 4.3   Joint equilibrium

one in Figure 4.2, can be treated as a series of connected subassemblies with fixed ends and designed for just two cases of gravity loading: (1) live and. dead load on two adjacent spans and (2) dead load on all spans and live load on alternate spans. The actual values of the design moments will depend upon the loads, spans, and relative member stiffnesses.

4.3   REVIEW OF FRAME MOMENT DIAGRAMS

Moment diagrams for frames are essentially the same as those for beams. The only difference is that with frames it is helpful to remember that the diagram must be compatible with the statics of joint equilibrium. This means that the signs of the various moments which act on the joints (moments provided by the members) must be such that the joint is in rotational equilibrium. The joints of Figure 4.3 all represent possible equilibrium conditions. In a four-member interior joint such as that of Figure 4.3*d*, the negative beam moments will oppose one another when the two spans are both loaded, and the moments in the columns will therefore be small (having only to sum to the difference between the two beam moments). However, when one of the spans is unloaded (alternate span loading pattern), both beam moments tend to rotate the joint with the same sense (Figure 4.3*e*). In that case the column moments will be significant.

If we can ignore the effects of sidesway or assume the frame is braced so that the joints remain essentially over and under one another, then a moment applied at the top of a column will result in a moment diagram that decreases linearly toward the bottom of the column. If the bottom is pinned, the moment will go to zero there. If the bottom is fully fixed, an analysis would show that the moment there will be exactly one-half the value of the moment applied at the top. This means a point of inflection will exist exactly one-third the way up the column. Figure 4.4

FIGURE 4.4

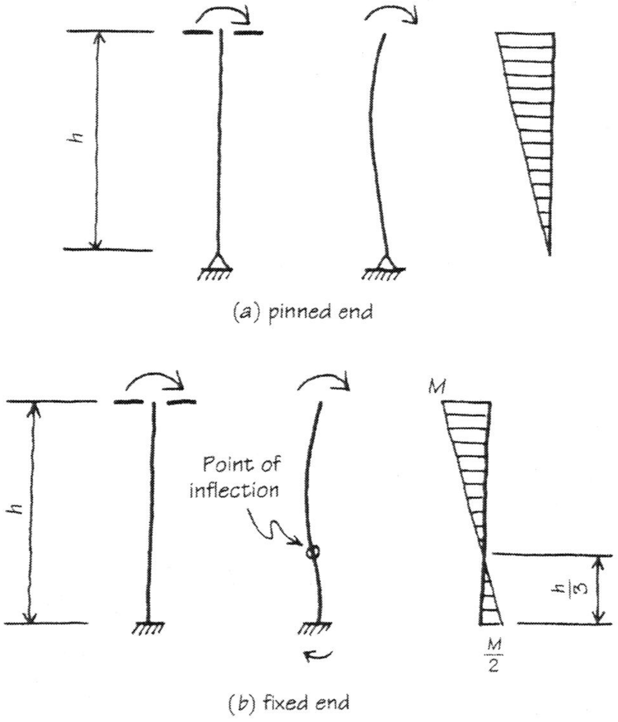

h

(a) pinned end

h

Point of inflection

M

$\frac{h}{3}$

$\frac{M}{2}$

(b) fixed end

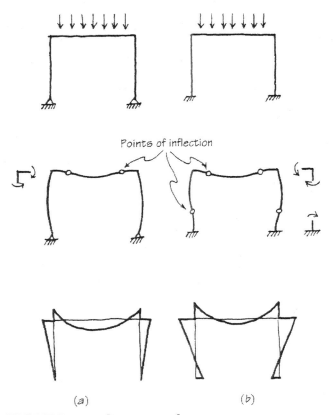

FIGURE 4.5 One-story bents.

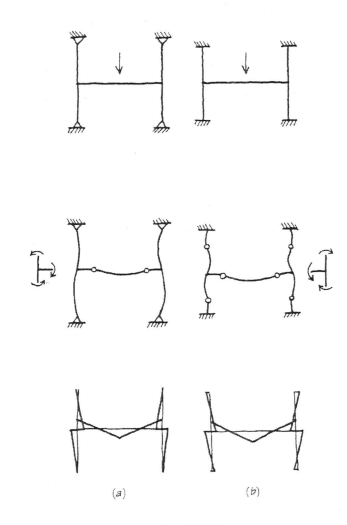

FIGURE 4.6 Point-loaded frames.

shows the deflected shapes and resulting moment diagrams for these two conditions. Figure 4.5 shows these same conditions on two frames that are identical except for the column base conditions. For both frames, note that the two moments at each corner are graphically equal in magnitude, as this is required for the corners to be in rotational equilibrium.

When three members make a joint (Figure 4.3c), one of the moments must act to oppose the rotation of the other two. Figure 4.6 illustrates this in a different manner on two similar frames. Notice that the moment diagram graphically shows two small moments from the columns which must sum to equal the larger moment value from the loaded beam. Notice also that the moments at the fixed ends in Figure 4.6b are graphi-

cally equal to one-half the value at the joint (or continuous) end.

Figure 4.7 shows a checkerboard loading pattern for a two-story frame. The reader should verify the joint sketches and how they relate to the deflected shape and moment diagram.

4.4 MOMENT ESTIMATING

Using the boundary conditions for a few "ideal" support situations, one can estimate

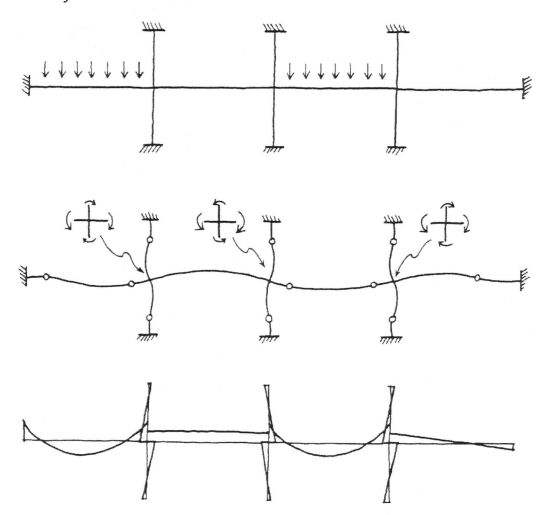

FIGURE 4.7

the magnitudes of the larger design moments in simple building frames with sufficient accuracy for preliminary design purposes. These estimated values can be obtained rapidly by making educated guesses about the relative amount of resistance to rotation provided at the ends of major members by the other attached members.

Basically, uniformly loaded beams in rigid frames are modified versions and combinations of the three beams of Figure 4.8, which represent support conditions that are *ideal* in that their ends are either fully fixed or fully pinned. Even with monolithic concrete construction, full fixity is almost impossible to achieve and the frictionless

pin or roller is well-known to exist in theory only. Thus, in reality the ends of members in continuous frames have degrees of restraint which must lie between fully fixed and fully pinned. A higher degree of end fixity means larger negative moments at the ends and a smaller positive moment at midspan (Figure 4.8a). A lower degree of end fixity means smaller negative moments and a larger positive moment (Figure 4.8c). A particularly large negative moment is achieved at one end when the other end is released and becomes free to rotate (Figure 4.8b). In fact, the left end moment is increased by exactly one-half of the moment that existed at the right end before it was released. Thus one-

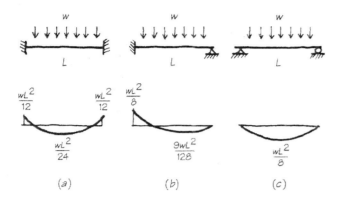

FIGURE 4.8

half of $wL^2/12$, which is $wL^2/24$, is added to the left end moment, making it $wL^2/8$.

Considering such a member as part of a continuous frame, we can say, in general, that the shorter (and therefore stiffer) and more numerous the members are that are attached monolithically to the end of the member, the larger the negative moment there will be. Figure 4.9 shows a progression of cases in which the same span has decreasing resistance to end rotation from Figure 4.9a to h due to changing numbers of restraining members and the effects of their end conditions. The end moments resulting from a uniform load on this span will diminish slowly from a value of $wL^2/12$ to a very small value, possibly in the order of $wL^2/35$ or $wL^2/40$.

Figure 4.10 shows a similar progression for the case of a beam with one end fixed and one end pinned. In this situation the moment at the left end due to a uniform load will decrease slowly from a high of $wL^2/8$ to a similarly small value.

Again using a uniform load, we can observe the effect that changing the restraint at one end has on the opposite end. Figure 4.11a shows a four-member frame with a

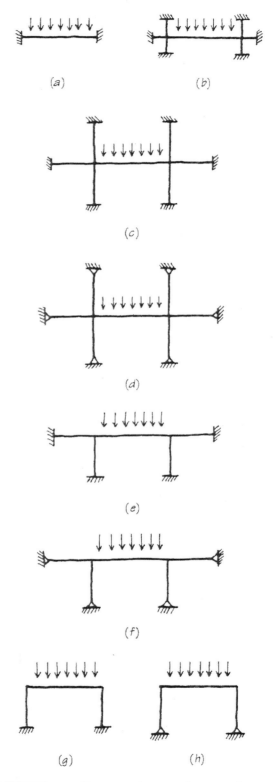

FIGURE 4.9 Decreasing end restraint.

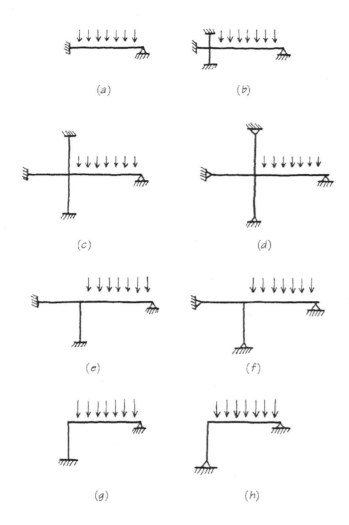

(a)   (b)

(c)   (d)

(e)   (f)

(g)   (h)

FIGURE 4.10   Decreasing end restraint.

moderately stiff left end. It will have a negative moment somewhat smaller than $wL^2/8$ [depending upon the stiffness (I/L) of the three adjoining members], say, about $wL^2/9$ or $wL^2/10$. Increasing the stiffness or restraint at the right end (moving from left to right in Figure 4.11) adds a negative moment to the right end and serves to reduce the left end curvature and moment. The moment diagrams and exaggerated curvature sketches reflect these changes. The left end negative moment in Figure 4.11c might be in the range of $wL^2/15$ or $wL^2/17$. All we can say for sure is that it is smaller than $wL^2/9$ or $wL^2/10$, which is the value we estimated for that same location when the right end was unrestrained (Figure 4.11a). The left end moment in Figure 4.11b would lie between the values that it would have for Figure 4.11a and c.

Figure 4.12 shows a more complete transition of end stiffnesses for a symmetrical frame. The values given are estimated negative end moments that would be applicable for either end. The kip-foot figures assume a load of 4 kips/ft over a 30-ft span. The actual values obtainable from an indeterminate analysis would, of course, depend upon the actual relative I and L values of the adjoining

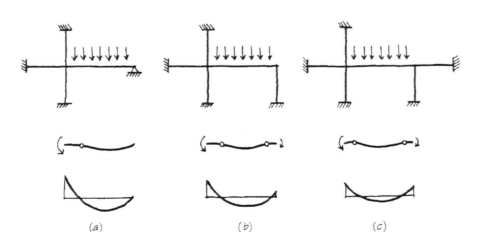

(a)   (b)   (c)

FIGURE 4.11
Increasing the stiffness of the right end adds negative moment at that point and decreases the opposite end moment *and* the positive moment.

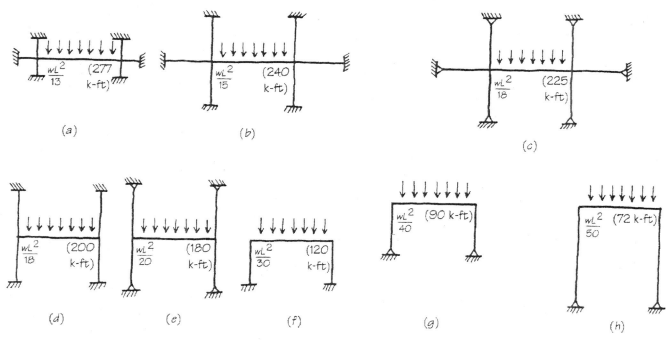

FIGURE 4.12   Possible end moment values assuming $w = 4$ kips/ft and $L = 30$ ft.

members.

It should be noted that when ends are more restrained and moment values are therefore larger, the difference between the moment values associated with any of the various restraints is less than when there is little stiffness and end moments are small. In the latter case a small change in end condition will cause a large change in the moment. This means that smaller end moments are harder to estimate accurately than

larger ones. The reduced accuracy is of little concern if one realizes that, in terms of preliminary design at least, only the larger moments are important for they will govern the overall member size.

Adding load to an adjacent span has the same effect upon the span under consideration as increasing the stiffness, only more so. Figure 4.13a shows a loaded girder whose negative moment at the left end (point x) is probably in the $wL^2/12$ to $wL^2/16$

FIGURE 4.13

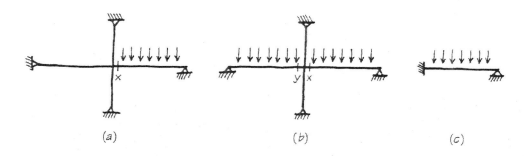

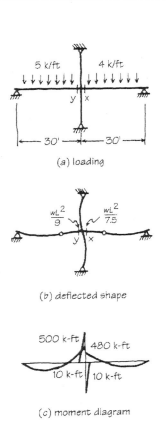

(a) loading

(b) deflected shape

(c) moment diagram

(d) joint equilibrium

FIGURE 4.14

range. This is rationalized as follows. The long member to the left offers very little restraint to rotation of the joint, and all three restraining members are pinned. This indicates the moment will be considerably less than $wL^2/8$ (the boundary condition moment that would exist there if the left end, point x, were fully fixed).

However, if the left span is loaded with the same amount of load as exists on the right span, then the moment at point x will shoot up to precisely $wL^2/8$, because now,

due to the balanced loading, the joint will not rotate at all and will act as though it were, indeed, fully restrained. The right-hand span will now behave exactly the same as the beam in Figure 4.13c. If we increased the load on the left-hand span even further so that it is larger than the load on the right-hand span, the joint will actually rotate slightly in a counterclockwise manner causing the moment at x to exceed the $wL^2/8$ value as it tries to oppose this rotation. In such a case the moment at y would be the larger of the two owing to the change in w values. This case is shown in Figure 4.14 using $L = 30$ ft and w values of 4 and 5 kips/ft for the right and left spans, respectively.

In this case the columns take very little moment because the almost equal loading on the beams causes very little rotation of the joint. Contrast this to the sketches in Figure 4.15 where the original frame of Figure 4.13a is analyzed. Here there is no doubt that the joint rotates clockwise under the influence of the load.

In this case the 257 kip-ft moment $(wL^2/14)$ would be a reasonable estimate and the moments placed upon the joint by the other members preserve its equilibrium. The 257 kip-ft moment could be divided up rather precisely if the I and L values were all known, but in their absence an equal split might be appropriate. This can be justified by noting that while the columns are short, the longer beam would have a larger I value; thus, the relative I/L values might be quite close.

As a general rule, when estimating moment values, estimate the larger, more critical negative moments first. The smaller ones are of less consequence and can often be obtained by comparing the loading and end conditions to those which generated the larger moments and by considering joint equilibrium.

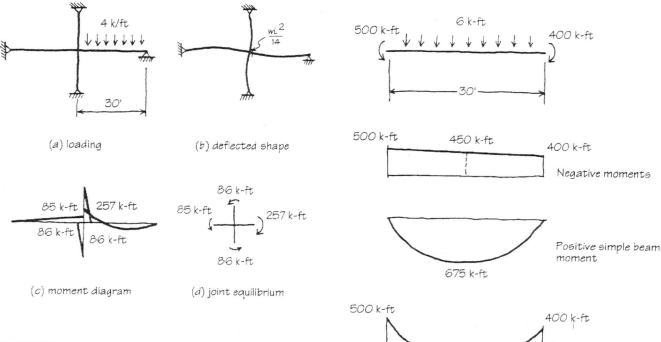

(a) loading   (b) deflected shape

(c) moment diagram   (d) joint equilibrium

FIGURE 4.15

FIGURE 4.16   Approximate determination of the positive moment value.

Positive moments should never be estimated; rather, they should be obtained from the member end moments via the shear diagram. For cases where the two end moments are not too different in terms of magnitude one can obtain the approximate positive moment value by averaging the end moments and algebraically adding the result to the simple beam midspan moment of $wL^2/8$. This is illustrated in Figure 4.16 using hypothetical values. It can be seen that the positive moment will only govern the member size when the end moment at one or both ends is small.

Frames with concentrated loads can be analyzed in a similar fashion by assuming a center span load location and basing estimates on the boundary values given in Figure 4.17.

Moment estimating is a valuable tool for designers, but, like any other tool, it takes

FIGURE 4.17

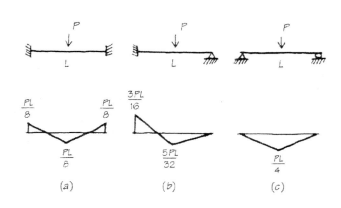

(a)   (b)   (c)

practice to become adept at its use. Experienced designers can usually guess the larger critical moment values within 10 or 15% of the actual values, more than close enough for preliminary design purposes. Once an approximate moment value is determined, the required cross-sectional dimensions to accommodate that moment safely and economically are found. Other chapters in this book are devoted to procedures and design aids that help achieve this next step in a straightforward manner.

4.5 ACI COEFFICIENTS

The American Concrete Institute (ACI) Code provides some coefficients which can be used in lieu of an indeterminate analysis for buildings of usual construction and commonly encountered load and span conditions. These are provided in Figure 4.18. In most instances they tend to be conservative, and therefore are more appropriate for preliminary design than for use in the final design process. They can be used for continuous slabs or beams, but the Code restricts their use to situations where:

1. Two or more spans exist.
2. Only uniform loads are present.
3. The service live load does not exceed three times the service dead load.
4. The longer of any two adjacent spans does not exceed the shorter by more than 20%.

In Figure 4.18 the span length L is to be taken as the *clear span* for the positive moment and the *average of two adjacent clear spans* for negative moments. When there are two different coefficients for negative moments on opposite sides of a supporting element, the larger moment should be used for both because only one pattern of negative moment steel will be placed across that

support when the design is finally executed.

For interior spans the maximum shearing force may be taken as half the span load, i.e., $wL/2$. For an exterior span, however, the stiffer interior support results in a larger negative moment. This increases the interior support reaction and decreases the exterior one. This is reflected in the 15% increase in maximum shear in Figure 4.18a.

The reader should notice a correlation between the moment coefficients in Figure

FIGURE 4.18   ACI coefficients.

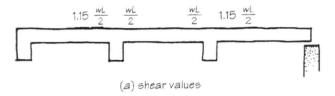

(a) shear values

Two spans

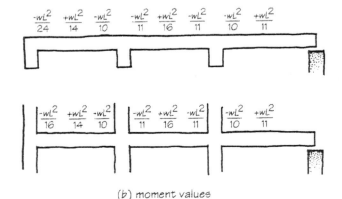

Three or more spans

(b) moment values

4.18*b* and the values for moments discussed in the previous section. They are based upon the same types of considerations and analyses. It should also be noticed that, in the case of continuous beams over several supports, the largest shears and moments will be located at the "exterior face of the first interior support." If one needs a "quick-and-dirty" maximum moment value without having to do any thinking, it is prudent to select $wL^2/8$ or $wL^2/9$. This will usually be conservative when applied to the longer spans that establish the structural and economic viability of a framing scheme. One must never forget, however, that a cantilever or overhanging beam will overrule all others with a determinate moment value of $wL^2/2$! For this reason and because of deflection problems, cantilever spans are usually quite short.

5

STRENGTH DESIGN CONCEPTS

5.1 HISTORY

The first theories concerning the behavior of reinforced concrete were developed in France and Germany in the 1880s. These early efforts evolved into what became known as the *working stress design* method. By 1910 this was a universally accepted approach to the design of reinforced concrete elements and remained so until the late 1960s. It has other names, frequently called *allowable stress design* or *service load design*. The underlying philosophy is that when the structure is carrying the maximum loads expected to act on it when it is "working" or in "service," certain stress levels in the concrete and the steel should not be exceeded. These are called the "allowable" stresses and are specified by groups such as the American Concrete Institute (ACI) as recommendations for code-making bodies.

In allowable stress design, classical formulas, such as that for bending stress, $f_b = Mc/I$, are used and the assumption of a *linear relationship between stress and strain* is made. The margin of safety is not specified numerically but exists inherently as the difference between the allowable stress and the failure stress. For years the ACI Code recommended that the allowable stress for concrete in compression be 45% of f_c' and that the allowable stress for steel in tension be 50% of f_y.

In the 1930s an American engineer, Charles S. Whitney, began his pioneering work in the development of *strength design*, the preferred design approach today and the one used almost exclusively. Whitney and others found considerable fault with the working stress method, principally with the fact that concrete does not have a linear stress-strain relationship. Also, since concrete crushes at a certain *strain* level, the factor of safety cannot be assured by the stipulation of an allowable stress. It was also felt that the working stress method was less suitable for concrete structures than for structures of timber and steel, not only because of the differences in stress-strain curves, but also because the method does not differentiate between dead and live loads. The magnitude of the dead loads in a building is known with considerably more accuracy than that of the live load, which is often rather arbitrarily stipulated by building codes according to the likely use of the space. Most designers would prefer a larger factor of safety for live loads due to their greater uncertainty, and this is not possible when using allowable stresses. In concrete structures the dead load is usually a relatively large fraction of the total load, and therefore the use of two different factors of safety, one for dead loads and the other for live loads, becomes desirable in terms of structural efficiency.

Whitney called the new approach the *ultimate strength design* method and based it upon the idea of calculating the theoretical failure capacity of a member and then making sure that the service loads (as defined below), *after being multiplied by appropriate factors of safety*, do not exceed that capacity. (This approach had long been used in the design of columns of any material; a failure or buckling load was computed and then divided by a factor of safety to achieve an allowable load.) No allowable stresses are needed and the fictitious assumption of a triangular stress pattern is avoided. Research showed that at failure the concrete stress distribution on the compression side of a beam took the shape of the concrete stress-strain curve of a test cylinder, and Whitney used this familiar shape as the basis for the theoretical calculations.

Acceptance of this new approach, later shortened in name to *strength design*, was delayed by World War II and not until 1956 was the method accepted by Committee 318

of the ACI. By 1977 the working stress method was relegated to a position in the appendix of the Code, having been supplanted by the new method since designers favored its more rational approach.

At this writing, in the U.S. and in most of the world, reinforced concrete structures are designed using the *strength design* method.

5.2 SERVICE LOADS

Service loads are the maximum loads that are expected to act upon the building while it serves its intended purpose. They consist of *dead loads* (the structural self-weight and all objects permanently attached to the structure such as partitions, ceilings, and all mechanical equipment), *live loads* (people, furniture, and all other movable objects), and *lateral loads* (from wind, earthquake, or retained soil).

Dead loads are relatively easy to compute or estimate from standard references and suppliers' literature. Live loads are almost always specified by local building codes which often choose to reference one of the three model codes in common use in the United States. Table 5.1 provides some representative uniform load values of service live loads for convenience, but the user is cautioned that local codes must govern the actual design in their individual jurisdictions.

5.3 LOAD FACTORS

The service loads are multiplied by factors of safety called *load factors*. The resulting amplified loads are called *ultimate loads* or sometimes (rather ambiguously) merely *design loads*. The process of member design then involves proportioning the member and its amount of reinforcing steel so that its ultimate capacity or "strength" is slightly

TABLE 5.1   *Representative Values of Uniform Service Live Loads*

Type of Space	Load (psf)
Hotels, dormitories, apartment buildings	
Private living areas	40
Public areas including corridors	100
Restaurants	100
Office buildings	
Office space	50
Public areas including corridors	100
Schools	
Classrooms	40
Corridors	80
Stores	
Retail	75
Wholesale	100
Theaters	
Fixed seats	50
Movable seats	100
Lobbies	100
Stands and bleachers	100
Libraries	
Reading areas	60
Stack areas	125+
Storage facilities	
Light	125
Heavy	250

greater than that needed to carry the ultimate loads. (As demonstrated by the examples of Chapter 7, this usually involves a first guess at the member's overall size so that the effect of self-weight can be included in the dead load.)

In the following discussion of load factors, some of the types of loads considered by the Code are not addressed. This is not because such loads are not important, but rather the fact that their inclusion would make this introductory text overly involved. These include loads due to temperature change, supported fluids, and those due to

soil and hydraulic soil pressure. (In Chapter 14, which treats retaining walls, soil pressure is treated as a live load, which has the same load factor as the Code-specified factor for such pressures.) The Code also requires that snow, rain, and other loads be treated differently because of statistical probability issues. This fact is ignored this book.

The Code provides that the required capacity be at least as great as

$$U = 1.2D + 1.6L \qquad (5\text{-}1)$$

where D = the dead load or its effects
L = the live load or its effects

Note that dead loads are assigned a lower load factor than live loads. This recognizes the fact that dead loads can be determined with greater accuracy than live loads. Live loads may vary considerably from the values specified by building codes and thus a higher factor of safety is warranted.

The load factors of 1.2 and 1.6 are usually applied directly to the loads but may be applied to the resulting moments or shears instead if this is more convenient.

In the unusual situation in which a member carries only dead load, the expression becomes

$$U = 1.4D \qquad (5\text{-}2)$$

This is to avoid an overall factor of safety of only 1.2.

Whenever lateral forces are involved, the Code recognizes that the worst case may actually exist under conditions of zero live load and maximum lateral load, where the gravity load acts in opposition to the lateral forces. Therefore, a check of the dead load plus lateral load combination is made taking the dead load factor as 0.9 rather than 1.2. (This ensures that the load factor does not contribute to a lesser degree of safety.) In any event the strength provided must never be less than that required by Equation (5-1), which will be used for all examples and problems of this book.

5.4 CAPACITY REDUCTION FACTORS

The Code also specifies a second but different kind of safety factor called the *capacity reduction factor*, a coefficient, always less than unity, that is used to reduce the calculated theoretical capacity of a member. It is given the symbol ϕ and Table 5.2 gives the various values for application to different situations. The ϕ factor is intended to take into account the relative importance of a member to the building's overall structural integrity and the relative uncertainties involved in design, construction, mode of failure, materials properties, and workmanship. The ϕ factor, for example, for a column is much smaller than for a beam because a column:

1. When it fails usually results in the collapse of a large portion of the building

2. Fails with little warning relative to a beam

3. Is more difficult to construct than a beam

The ϕ factor therefore gives us a larger overall factor of safety for columns than for beams. The actual factor of safety at work in strength design is

$$F.S.= \frac{\text{load factors}}{\phi \text{ factor}}$$

and it ranges from 1.33 to 2.46 depending upon the types of load involved and the member being designed.

TABLE 5.2 Capacity Reduction (ϕ) Factors

Type of Stress	ϕ
Bending	0.90
Shear and torsion	0.75
Compression (column with spiral tie reinforcing)	0.70
Compression (column with lateral tie reinforcing)	0.65
Compression (bearing)	0.65

6

FLEXURE THEORY

6.1 THE INTERNAL MOMENT COUPLE

The typical expression for allowable moment capacity in working stress theory is

$$M = SF_b \qquad (6\text{-}1)$$

where S = elastic section modulus (in³)
F_b = allowable bending stress (psi or ksi)

S is equivalent to I/c, where c is the extreme fiber distance. Therefore, for a rectangular section

$$S = \frac{bd^3/12}{d/2}$$

$$S = \frac{bd^2}{6}$$

and thus

$$M = \frac{bd^2}{6} F_b \qquad (6\text{-}1a)$$

This same expression can be obtained by examining the triangular stress distribution in such a beam. Figure 6.1 shows a rectangular beam section of width b and depth d acted upon by a positive moment couple. The compressive and tensile stress distributions will be equal by virtue of symmetry and, of course, T and C must be equal to meet the requirements of horizontal equilibrium; i.e., $\Sigma F_x = 0$. Because the centroid of a triangle is located at its third point, it is easy to see that the moment arm of the internal moment couple is two-thirds the depth in length. The magnitude of either force resultant is equal to the volume of one of the stress triangles:

$$C = T = \frac{1}{2} F_b \left(\frac{d}{2}\right)(b) = \frac{bd}{4} F_b$$

Since the magnitude of a couple is the product

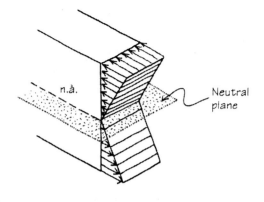

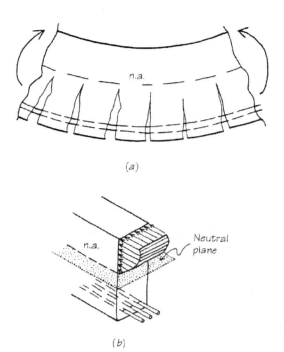

(a)

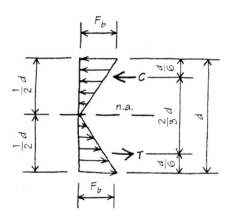

FIGURE 6.1   Elastic stresses in a rectangular section.

of one of the forces times the distance between the two forces (moment arm), we get

$$M = C\left(\frac{2}{3}d\right) = T\left(\frac{2}{3}d\right)$$

$$M = \frac{bd}{4}(F_b)\left(\frac{2}{3}d\right)$$

$$\boldsymbol{M = \frac{bd^2}{6}F_b} \qquad (6\text{-}1a)$$

6.2   DEVELOPMENT OF THE RESISTING MOMENT EQUATION

In reinforced concrete we can develop the same type of expression for the internal mo-

(b)

FIGURE 6.2

ment capacity. Using strength design rather than working stresses, we find that the stresses acting on the cross section will be those present under failure load conditions. The assumption is made that the concrete is cracked below the neutral axis (Figure 6.2a), but since it is firmly bonded to the steel, the tensile stresses are transferred so that the entire tensile force resultant is carried by the steel. (The assumption that tensile cracks propagate all the way to the neutral axis is not quite true because concrete does have some tensile strength, but it is standard practice to ignore this and the assumption is conservative. As discussed in Section 2.8, it is very important to realize that the carrying capacity of any reinforced concrete beam is totally dependent upon the development of proper bond between the concrete and the tensile reinforcement.) Because of the cracks, the section is not symmetrical and the neutral axis is not at middepth. Rather, it is at the bottom of the compressive stress distribution as shown in Figure 6.2b. As a concrete beam is loaded the compressive and tensile force resultants increase just as in a timber or steel section, but the moment arm also increases as the tensile cracks propagate upward changing the location of the neutral axis. Figure 6.3 shows the stress distributions under increasing load. It is the situation shown in Figure 6.3c that is important to the development of a strength design formula for ultimate moment capacity.

When the beam fails, the stress in the tensile steel will always be at yield. (This is assured by a Code provision and is discussed in Section 6.4.) Once the steel yields, the tensile force resultant T becomes constant. (Strain hardening, which normally takes place in mild steel, is ignored because of its uncertain magnitude.) Because of horizontal equilibrium, C is equal to T and therefore is also fixed in value. Any further increase in moment capacity is limited to the small increase in moment arm, which takes place as the crack continues to propagate upward under the increasing strains. This change in moment arm is very limited because the area of concrete remaining for the compressive stresses to act upon is decreasing and with C held constant the compressive stresses will build up rapidly. The strain is maximum at the edge of the beam- and decreases linearly to zero at the neutral axis. Final collapse of the beam takes place when the concrete strain reaches approximately 0.003.

It is no accident that the compressive stress distribution at failure looks like the stress vs. strain curves of Section 2.1; the behavior of concrete when it crushes is very much the same whether it is in a test cylinder or part of a beam.

From Figure 6.4a, the magnitude of the internal moment couple is

$$M = C \times \text{(moment arm)}$$

or $M = T \times \text{(moment arm)}$

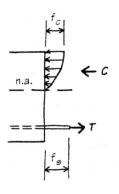

(a) moderate load

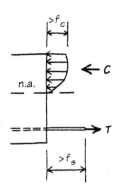

(b) increased load

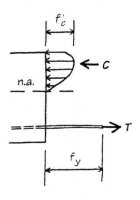

(c) failure load

FIGURE 6.3 Stresses in a reinforced concrete beam as the load increases.

The magnitude of C is somewhat difficult to determine because of the curved shape of the compressive stress pattern; however, the

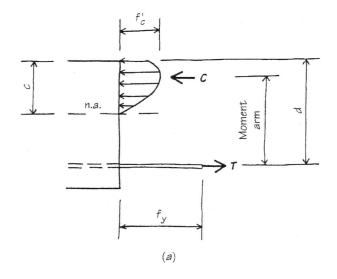

(a)

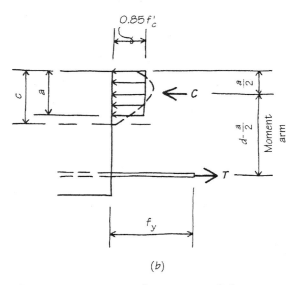

(b)

FIGURE 6.4   Development of the equivalent stress block.

tensile force resultant is easy to find because at yield

$$T = A_s f_y \qquad (6\text{-}2a)$$

where A_s = the area of the tensile steel.

The finding of the magnitude of C and the consequent determination of the size of the moment arm is greatly simplified by the concept of the *equivalent rectangular stress block* shown in Figure 6.4*b*. This fictitious

stress pattern is proportioned such that the force resultant of the rectangular block will have the same magnitude and location as that of the actual nonlinear stress volume. These proportions were determined experimentally by Charles S. Whitney (see Section 5.1) in the early 1930s and are given as a uniform stress level of 0.85 f'_c and a reduced height of dimension a. The difference between a and c in Figure 6.4*b* accounts for the skewed shape of the real stress distribution. As indicated in Section 2.1, the curve becomes more sharply skewed as concrete strength increases. According to the American Concrete Institute (ACI) Code, the relationship between a and c can be taken as

$$a = 0.85c \qquad (6\text{-}3)$$

up to an f'_c value of 4000 psi. Thereafter, the 0.85 factor must decrease by 0.05 for each 1000 psi increase in f'_c. However it need never be taken as less than 0.65. This relationship is graphed in Figure 6.5. (From a practical standpoint, the relationship between a and c need not be determined for most purposes. It is included here in the interest of completeness in the development of strength design theory.)

The compressive force resultant for a beam of width b is now readily stated as

$$C = 0.85 \, f'_c \, ab \qquad (6\text{-}2b)$$

FIGURE 6.5

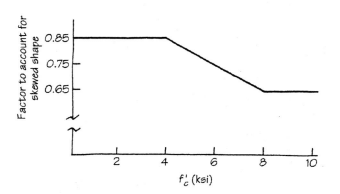

and the moment arm (see Figure 6.4*b*) is easily dimensioned as

$$d - \frac{a}{2}$$

where $d =$ the effective beam depth (This is defined as the distance from the extreme compressive fiber to the centroid of the tensile steel.)

The internal moment couple is then

$$M = C\left(d - \frac{a}{2}\right)$$

or

$$M = T\left(d - \frac{a}{2}\right)$$

These can be restated as

$$M = 0.85 f_c' ab\left(d - \frac{a}{2}\right) \qquad (6\text{-}4a)$$

or

$$M = A_s f_y\left(d - \frac{a}{2}\right) \qquad (6\text{-}4b)$$

Assuming that one knows the beam dimensions, the strengths of the materials, and the amount of steel, the only unknown is a, the depth of the fictitious stress block. The value of a can be established in terms of the known values by remembering that C must always equal T:

$$0.85 f_c' ab = A_s f_y$$

or

$$a = \frac{A_s f_y}{0.85 f_c' b} \qquad (6\text{-}5)$$

We can substitute into either of the Equations (6-4) to find the moment capacity of a beam. Equation (6-4*b*) is simpler and we get

$$M = A_s f_y\left(d - \frac{A_s f_y}{2(0.85 f_c')b}\right)$$

or

$$M = A_s f_y\left(d - 0.59\frac{A_s f_y}{f_c' b}\right) \qquad (6\text{-}6)$$

This is a valid expression for the theoretical failure moment of a rectangular beam with only tensile reinforcement; i.e., it ignores any reinforcement in the compressive part of the cross section.

Most designers prefer to express A_s in the equation by the *reinforcement ratio*, defined as

$$\rho = \frac{A_s}{bd} \qquad (6\text{-}7)$$

After substituting ρbd for A_s in both places and simplifying, we get

$$M = \rho f_y bd^2\left(1 - 0.59\rho\frac{f_y}{f_c'}\right) \qquad (6\text{-}6a)$$

If we include ϕ, the capacity reduction factor, as addressed in Section 5.4 and mandated by the Code, we get the most recognized equation for the *resisting moment*:

$$\boxed{M_r = \phi\rho f_y bd^2\left(1 - 0.59\rho\frac{f_y}{f_c'}\right)} \qquad (6\text{-}8)$$

Equation (6-8) looks complicated, but the reader should realize that it is nothing more than

$$\boxed{M_r = \phi T\left(d - \frac{a}{2}\right)} \qquad (6\text{-}8a)$$

In the analysis of a rectangular beam using only tensile reinforcement, it is merely necessary to ensure that the value of M_r is always equal to or larger than M_u, the factored load moment.

Example 6.1 The simple beam of Figure 6.6 is loaded by a concentrated service live load at midspan. It has the cross section shown; if $f_c' = 3000$ psi and $f_y = 60$ ksi, will the section be adequate?

Solution: First determine the self-weight and then find the factored load moment. Compare the factored load moment to the resisting moment for the section.

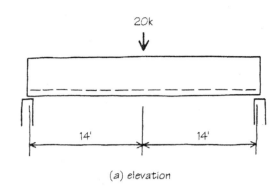

(a) elevation

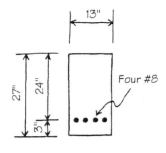

(b) midspan section

FIGURE 6.6

The overall dimensions are 13×27 in and the unit weight of reinforced concrete is 150 pcf. Thus, the self-weight per unit length is

$$\frac{13(27)}{144}(150) = 366 \text{ plf}$$

Applying the load factor for dead loads as explained in Section 5.3, we get

$$w_u = 1.2(366) = 439 \text{ plf} \approx 0.44 \text{ kip/ft}$$

The concentrated live load is also factored to get
$$P_u = 1.6(20) = 32 \text{ kips}$$

The ultimate moment exists at midspan:

$$M_u = \frac{w_u L^2}{8} + \frac{P_u L}{4}$$
$$= \frac{0.44(28)^2}{8} + \frac{32(28)}{4}$$
$$= 43 + 224$$
$$= 267 \text{ kip-ft}$$

To find the resisting moment, first determine the reinforcement ratio. From Table A.1 of the Appendix, the area of four #8 bars is 3.16 in²:

$$\rho = \frac{A_s}{bd} \qquad (6\text{-}7)$$
$$= \frac{3.16}{13(24)}$$
$$= 0.010$$

Then

$$M_r = \phi \rho f_y bd^2 \left(1 - 0.59 \rho \frac{f_y}{f_c'} \right) \qquad (6\text{-}8)$$

From section 5.5, ϕ for bending is 0.9, thus
$$M_r = 0.9(0.010)(60)(13)(24)^2 \left[1 - 0.59(0.010)\frac{60}{3} \right]$$
$$= 3570 \text{ kip-in}$$
$$= 297 \text{ kip-ft}$$

Since $M_r > M_u$, the section is adequate. Furthermore, since M_r is only about 10% greater than M_u, the beam section is properly designed in terms of structural efficiency and economy.

Example 6.2 Determine the resisting moment for the cantilever beam of Figure 6.7 and the amount of total uniform load per foot, w_u, that would result from this moment. Let $f_c' = 4000$ psi and $f_y = 60$ ksi.

Solution: In this case the rectangular stress block will be on the bottom as in Figure 6.8, but the resisting moment equation remains the same. Determine M_r and then establish w_u by knowing that the maximum moment is $w_u L^2/2$.
$A_S = 8.0$ in² and the reinforcement ratio is

$$\rho = \frac{A_s}{bd} \qquad (6\text{-}7)$$
$$= \frac{8}{14(33)}$$
$$= 0.017$$

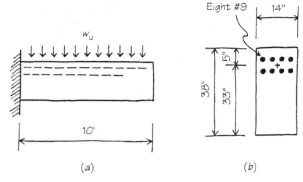

(a)   (b)

FIGURE 6.7

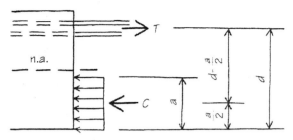

FIGURE 6.8   Internal moment couple for Example 6.2

$$M_r = \phi \rho f_y b d^2 \left(1 - 0.59\rho \frac{f_y}{f_c'}\right) \qquad (6\text{-}8)$$

$$= 0.9(0.017)(60)(14)(33)^2 \left[1 - 0.59(0.017)\frac{60}{4}\right]$$

$$= 12\,100 \text{ kip-in}$$

$$= 1000 \text{ kip-ft}$$

The amount of w_u that would generate this M_r is

$$w_u = \frac{2M_r}{L^2}$$

$$= \frac{2(1000)}{(10)^2}$$

$$= 20 \text{ kips/ft}$$

This amount would include all the factored live and dead loads, including self-weight.

6.3   MINIMUM REINFORCEMENT RATIO

The code specifies that beams shall not have less steel than

$$\rho_{\min} = \frac{200}{f_y}$$

The purpose of this provision is to ensure that

as a tensile crack develops there will be enough steel at least to carry the tensile forces that the concrete was carrying before it cracked. This requirement seldom controls except in beams made very large for architectural reasons. The Code permits the minimum requirement to be waived if the amount of steel used is one-third larger than that required by M_u at the various beam sections.

The minimum steel requirement was not at issue in Examples 6.1 and 6.2, as ρ was well above the required value in each case.

6.4   MAXIMUM REINFORCEMENT RATIO

From a conceptual standpoint, if the percentage of tensile steel is small, the steel stress will be high and surely will go to yield before the concrete stress gets very large. Only after the steel yields and cracks propagate toward the compression zone, resulting in large beam curvatures and deflections, will the concrete stresses get large. From a safety standpoint, this condition is very desirable because a slowly deforming beam provides a lot of warning before it collapses. On the other hand, a beam with a large percentage of steel will have very low steel stresses and the concrete might actually crush without having the steel reach its yield value. Such a situation is clearly dangerous; if the beam fails by crushing of the concrete before there is much strain in the steel, the failure will exhibit little

FIGURE 6.9
Laboratory test of an underreinforced beam. It maintains a large moment capacity while deflecting, giving warning of an impending collapse. (*Professor William L. Gamble, Department of Civil Engineering, University of Illinois at Urbana-Champaign*)

or no warning.

It is logical that there is some intermediate amount of steel that would result in a simultaneous yielding of the steel and crushing of the concrete. This particular ratio is called the *balanced steel ratio* and is given the symbol ρ_b. It can be obtained for any combination of steel and concrete strengths by examining the strain picture of the beam at failure. Beams which have less steel than ρ_b are called *underreinforced*, and those in which the amount of steel exceeds ρ_b are called *overreinforced*. The Code mandates a *maximum permissible steel ratio 25% less than* ρ_b to guarantee an underreinforced or ductile, large deformation failure under overload conditions. This is shown in the laboratory test beam of Figure 6.9. A beam with a lot of steel may be strong but it is certainly not safe.

$$\rho_{max} = 0.75\rho_b$$

Table A.2 of the Appendix provides ρ_b and ρ_{max} values for selected combinations of concrete and steel strengths. Checking that table, we find that the reinforcement ratios for Examples 6.1 and 6.2 were both well below ρ_{max} (0.010 < 0.0160 and 0.017 < 0.0213, respectively).

The percentage of steel to use in a beam is up to the designer and the range of values between ρ_{min} and ρ_{max} is quite large. Most beams have moderate amounts of steel in them for the following reasons. A small amount of steel will require a very large beam wasteful of both space, which is usually at a premium, and concrete. A large amount of steel will provide for a smaller beam cross section, of course, but may be very uneconomical in terms of the high cost of steel and the difficulty of placing the concrete around the closely spaced reinforcing bars. A smaller beam section may also encounter deflection problems due to reduced section properties. For preliminary design purposes we prefer to keep ρ in the 40 to 60% of ρ_b range, i.e., about two-thirds of ρ_{max}. This will result in economical yet not too large cross sections.

PROBLEMS

6.1 If the beam of Example 6.1 were reinforced with four #9 bars, could it safely carry an additional service live load of 1 kip/ft?

6.2 Find the resisting moment of a 16 × 33-in rectangular beam section reinforced for tension with six #8 bars. The effective depth is 30 in, f_y = 60 ksi, and f'_c = 3500 psi.

6.3 If a simply supported beam of the section given in Problem 6.2 were to span 33 ft, what is the greatest imposed uniform service load (w) it could support?

6.4 A rectangular beam section 14 in wide with an effective depth of 24 in is reinforced with eight #10 bars (f_y = 60 ksi). If the effective depth of the beam were increased to 30 in and the area of tensile steel reduced to 7.62 in^2, would the altered section be more or less moment resistant than the original? Let f'_c = 4000 psi.

6.5 What is the maximum compressive force resultant that could develop in a beam of 13-in width and 28-in effective depth, where f_y = 60 000 psi and ρ = 0.0100?

6.6 A simply supported beam 22 ft long is reinforced for tension with six #9 bars. Factored loads (including self-weight) total 6 kips/ft. f_y = 60 ksi, and f'_c = 4500 psi; if b = 11 in and d = 20 in, is the section acceptable?

6.7 The only load to be imposed on a simply supported beam of 18-in width, 38-in effective depth, and 44-in overall depth spanning 40 ft is a concentrated dead load at midspan. How great a concentrated service dead load can the beam safely support at this location if f'_c = 4000 psi and the 60-ksi tension reinforcement is designed to a ratio of ⅔ρ_{max}?

7

BEAM DESIGN FOR FLEXURE

7.1 BASIC DESIGN PROCEDURE FOR A SINGLY REINFORCED BEAM

A singly reinforced beam is one that has been analyzed or designed taking into account only the reinforcement in the tensile zone. Reinforcement that exists in the compression zone to increase the strength of the beam in compression (Section 7.4) or for other reasons such as ease of construction or deflection control (Chapter 9) is ignored in this procedure.

Equation (6-8) can be transposed somewhat to make it more useful for design purposes by grouping the dimensional factors and the materials factors:

$$\frac{M_r}{\phi bd^2} = \rho f_y \left(1 - 0.59\rho \frac{f_y}{f_c} \right) \qquad (7\text{-}1)$$

If we let the right-hand side of the equation take the value of R, then an expression for d can be more readily developed:

$$M_r = \phi bd^2 R \qquad (7\text{-}2a)$$

$$\boxed{d = \sqrt{\frac{M_r}{\phi bR}}} \qquad (7\text{-}2b)$$

For a first trial at getting a required d value, the designer selects:

f_c' and f_y (both often constant for a large percentage of the beams on a project)

b (usually established as the same as the column width for ease of construction)

ρ (the most influential parameter)

Example 7.1   Assuming that $b = 15$ in, $f_c' = 3000$ psi, $f_y = 60$ ksi, and $\rho = \frac{2}{3}\rho_{max}$, determine the d needed to carry an M_u value of 350 kip-ft. That is, M_r must be at least as great as 350 kip-ft.

Solution: From Table A.2, ρ_{max} is given as 0.0160. Thus, $\frac{2}{3}\rho_{max} = 0.0107$. Using the given values, we calculate

$$R = \rho f_y \left(1 - 0.59\rho \frac{f_y}{f_c}\right)$$

$$= 0.0107(60)\left[1 - 0.59(0.0107)\frac{60}{3}\right]$$

$$= 0.561 \text{ ksi}$$

[Note that R has units of ksi even though it is not a stress. It is important to realize that Equations (7-2) must be *consistent* in terms of *units*.] Solving for d, we get

$$d = \sqrt{\frac{M_r}{\phi b R}} \qquad (7\text{-}2b)$$

$$= \sqrt{\frac{350(12)}{0.9(15)(0.561)}}$$

$$= 23.5 \approx 24 \text{ in}$$

Tables B.1 of the Appendix provide the R values for three different concrete strengths and the full range of ρ values. Entering Table B.1(3) for $f_c' = 3000$ psi concrete with a ρ value of 0.0107, we can verify our R value of 0.561 ksi.

Example 7.2   Use the same data presented for Example 7.1 except let $\rho = \rho_{max}$. Determine the required effective depth.

Solution: Note that by setting $\rho = \rho_{max}$, we shall obtain the *minimum* permissible d (for a beam with tension reinforcement only) using these materials. From Table A.2, ρ will be 0.0160. Then from Table B.1(3) we get $R = 0.779$ ksi.

Then

$$d = \sqrt{\frac{M_r}{\phi b R}} \qquad (7\text{-}2b)$$

$$= \sqrt{\frac{350(12)}{0.9(15)(0.779)}}$$

$$= 20 \text{ in}$$

It should be noted that such a shallow depth may not be viable in terms of shear and deflection, but here we are concerned only with the moment requirement. In the interest of structural efficiency many designers control the depth-to-width ratio of rectangular beams. A common rule of thumb keeps this ratio between 1.5 and 3.0, but this is not a hard and fast range. It is true that shallow beams have deflection problems and are not economical in terms of the use of materials. It is also true that very deep narrow beams have stability problems and may be difficult to construct. For example, the concrete in zones of negative moment has to be placed through several layers of reinforcing bars if the beam is narrow.

Tables B.1 can also be used to determine the steel ratio needed for a given depth. This is a common situation in design where the maximum moment (e.g., the negative moment at the end of a beam in a continuous frame) has established the depth to be used throughout the beam length. The smaller positive moment out in the middle of the span will require a smaller amount of steel. The next example illustrates this determination.

Example 7.3   Again using the data presented in Example 7.1 and the d obtained of 24 in, determine the ρ value required at midspan if the moment there is 225 kip-ft.

Solution: Using the known moment and depth, solve for R and enter Table B.1(3) to find the needed ρ:

$$R = \frac{M_u}{\phi bd^2} \tag{7-2a}$$
$$= \frac{225(12)}{0.9(15)(24)^2}$$
$$= 0.347 \text{ ksi}$$

Using Table B.1(3), the steel ratio needed will be

$$\rho = 0.0063$$

7.2 USE OF PRELIMINARY DESIGN GRAPHS

The Graphs C.1 in the Appendix enable rapid determination of effective depth for preliminary design purposes. Curves are plotted for a range of width values for each of three different concrete strengths. A steel yield value of 60 ksi and a ρ value of $\frac{2}{3}\rho_{max}$ are used throughout. Knowing the M_r required and selecting a width, we can obtain d directly. These graphs are useful for getting a rough first approximation of the needed effective depth. It can later be refined by modifying ρ as needed and using Tables B.1.

Example 7.4 The maximum moment in the continuous beam of Figure 7.1 occurs at the ends and has been estimated to be

$$M_u = \frac{P_u L}{9} + \frac{w_u L^2}{13}$$

The service loads are $P_{ll} = 10$ kips, $P_{dl} = 12$ kips, $w_{ll} = 2$ kips/ft, and $w_{dl} = 2$ kips/ft. In this case the dead load includes the beam self-weight. Assuming $f'_c = 3000$ psi, $f_y = 60$ ksi, and a beam width of 14 in, determine the required approximate effective depth.

Solution: Let $\rho = \frac{2}{3}\rho_{max}$ so we can use Graph C.1(3). The ultimate loads will be

$$P_u = 1.2(12) + 1.6(10)$$
$$= 30.4 \text{ kips}$$
$$w_u = 1.2(2) + 1.6(2)$$
$$= 5.6 \text{ kips/ft}$$

The design moment will then be

$$M_u = \frac{30.4(38)}{9} + \frac{5.6(38)^2}{13}$$
$$= 128 + 622$$
$$= 750 \text{ kip-ft}$$

FIGURE 7.1 Figure for Example 7.4.

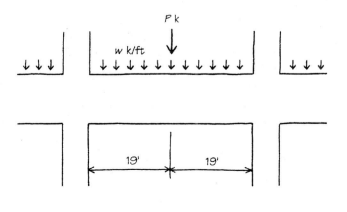

FIGURE 7.2 Cantilever for Example 7.5.

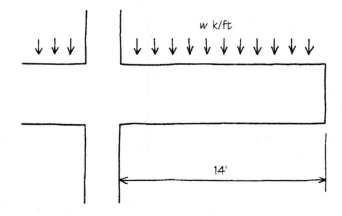

Entering Graph C.1(3) with 750 kip-ft gives us a preliminary effective depth of about 36.0 in. The overall depth will, of course, be larger due to the placement pattern of the steel, stirrups, and the required 1½-in cover.

Example 7.5   Determine the approximate required depth of the cantilever beam in Figure 7.2 given the following data: $w_{ll} = 2.5$ kips/ft, $w_{dl} = 1.2$ kips/ft, $f'_c = 4000$ psi, $f_y = 60$ ksi, and $b = 16$ in.

Solution: First estimate the beam's overall depth so that self-weight may be added to the dead load. Guessing how deep a beam will be is a matter of experience. Generally speaking, cantilevers are usually quite deep; in this case let's try 32 in. If this is far off, it will be relatively simple to modify and cycle through the process again. The self-weight per foot is easy to find by multiplying the number of square feet in the cross section by 150 pcf.

$$\text{s.w.} = \frac{16(32)}{144}(150)$$

$$= 533 \text{ plf}\quad\text{or } 0.53 \text{ kip/ft}$$

The ultimate load will then be

$$w_u = 1.2(1.2 + 0.53) + 1.6(2.5)$$

$$= 6.1 \text{ kips/ft}$$

which gives us a design moment of

$$M_u = \frac{w_u L^2}{2}$$

$$= \frac{6.1(14)^2}{2}$$

$$= 598 \text{ kip-ft}$$

Assuming $\rho = \frac{2}{3}\rho_{max}$ enables us to use the graphs. Entering Graph C.1(4) with 598 kip-ft, we get a preliminary effective depth of about 26 in. The

assumed overall depth or thickness was 32 in; this is more than adequate for the reinforcing steel and the required cover. The thickness could probably be reduced to 29 or 30 in.

Example 7.6   Assume the cantilever of the previous example is now 21 ft long. Determine the approximate depth needed for the moment if other conditions remain the same.

Solution: Assuming a larger overall depth of 48 in, we get a new self-weight estimate of

$$\text{s.w.} = \frac{16(32)}{144}(150)$$

$$= 800 \text{ plf}\quad\text{or } 0.80 \text{ kip/ft}$$

The ultimate load will be

$$w_u = 1.2(1.2 + 0.80) + 1.6(2.5)$$

$$= 6.4 \text{ kips/ft}$$

and the new design moment is

$$M_u = \frac{6.4(21)^2}{2}$$

$$= 1410 \text{ kip-ft}$$

Graph C.1(4) will then give us an approximate required depth of 40 in. Again, we overestimated the required thickness. However, a beam of this size would probably need two or even three layers of reinforcing, so the beam could only get two or three inches smaller.

It should be noted that the depth of 48 in while sufficient for moment will not be adequate to control deflection. This beam is examined for deflection in Example 9.2 of Chapter 9 and found to have a severe problem. The reader should be cautioned that just as with timber and steel, long cantilevers are usually controlled by deflection.

7.3 PLACEMENT OF REINFORCE-MENT

The American Concrete Institute (ACI) Code specifies the minimum spacing requirements of reinforcing bars in order to ensure an adequate bond between the concrete and steel and to facilitate the ease of construction. These requirements were addressed in Section 2.9.

Table A.3 of the Appendix provides the minimum width values necessary to accommodate various numbers of bars in a single layer. The table assumes a ¾-in maximum aggregate size, #4 stirrups, and 1½ in of cover on each side. Using Tables A.1 and A.3 we can easily select the required steel for a beam. (In monolithic slab and beam construction the outer bars of the top layer of negative moment steel may actually be in the adjacent slab area.)

Example 7.7 Select the steel required for the beam of Example 7.1.

Solution: With a ρ value of 0.0107, a beam width of 15 in, and the effective depth determined at 24 in, the amount of steel needed is

$$A_s = \rho bd$$

$$= 0.0107(15)(24)$$

$$= 3.85 \text{ in}^2$$

Using Table A.1, we find that enough reinforcement is provided by any of the following choices:

Three #11, $A_s = 4.68 \text{ in}^2$

Four #9, $A_s = 4.00 \text{ in}^2$

Five #8, $A_s = 3.93 \text{ in}^2$

Seven #7, $A_s = 4.20 \text{ in}^2$

Using Table A.3, we find that all but the last choice will fit within the 15-in width. The second or third selection would work very well; the first choice would provide too much steel and could also have bond or crack control problems. It is usually best to avoid a small number of large diameter bars in the typical section. The distribution of steel over the tensile zone of a beam is governed by the Code in the interest of controlling the width of cracks that develop. This is important to maintain the effectiveness of the cover requirements discussed in Section 2.9.

More than one size of bar may be used in a pattern in order to provide an amount of steel greater than, but reasonably close to, that required, provided that symmetry is maintained. As a general rule construction is made more straightforward if bar sizes are not mixed; but if they are, try to use only two sizes reasonably close together, e.g., #8s and #10s. Bars only one size apart are hard to distinguish from one another and bars that are many sizes apart can make for an uneven distribution of tensile forces and bond stresses.

More information regarding the placement of reinforcing may be found in Section 2.9.

7.4 DOUBLY REINFORCED BEAMS

Some beams require the use of steel in the compression zone in order to develop the necessary moment capacity. This technique is available when a very shallow beam must be used because of restricted headroom space. The use of compression steel increases the compressive force resultant enabling us to increase the amount of tensile steel and still maintain the required *underreinforced* behavior. It should be noted that doubly reinforced beams are generally avoided due to cost and elastic deflection requirements.

Compression steel can, however, be used to advantage if creep is a concern (Chapter 9). In this case steel is added to decrease the concrete compressive stress level, thus reducing the tendency for the concrete to creep.

7.5   T BEAM BEHAVIOR

The monolithic nature of most cast-in-place construction means that the floor slab and beam can act as one unit in the shape of a *T*. The top or slab portion of the *T* is called the flange and the part projecting downward is called the stem. For most beams of continuous construction the negative end moments are the controlling design values and the flange of the T beam will be in tension. Apart from some requirements regarding the distribution of the tensile steel, the beam is designed the same as a simple rectangular beam having a width equal to that of the stem. In the zones of positive moment, however, advantage can be taken of the large width of the beam on the compression side. Code stipulations dictate the width to be used for design in such cases.

7.6   DESIGN AIDS

A number of handbooks of tables and graphs are available to speed up the design process. Most of them provide material similar to that contained in the Appendix, except in more complete formats. The American Concrete Institute publishes a useful book of design aids called the *Design Handbook*. It is a two volume compendium of tables and graphs for footings, beams, slabs, and columns and is quite thorough in terms of its treatment of reinforcing selection and location. Another popular book is the *CRSI Handbook* published by the Concrete Reinforcing Steel Institute. It is thorough in its coverage of columns and is especially useful in the design of floor systems of all types. It also has sections on retaining walls, footings, and piles.

In recent years handbooks have been supplanted by computer programs which are very helpful during the final design process. Numerous companies sell software for almost any type of mini- or microcomputer, and most engineering offices design concrete elements only with the aid of a computer. The output is checked by hand calculations or with the help of tables and graphs. Some of the available programs will generate sophisticated detail drawings as well as establish member proportions and steel requirements. Many engineering firms have developed in-house software to meet their own specifications and ways of doing things.

At this writing most of the computer-aided design software in use is not very appropriate for preliminary design or structural planning purposes. It is probable that good interactive preliminary design programs will become available in the near future. *It is not likely, however, that the highly complex task of integrating spatial organization and structural planning with other architectural and engineering requirements will ever be accomplished except through the use of human experience and judgment.*

PROBLEMS

7.1 A rectangular beam of 16-in width carries a maximum factored moment, which is positive, of 592 kip-ft; $f_y = 60$ ksi and $f_c' = 5000$ psi.
    (a) Find the effective depth required for this beam if ρ is set at 0.0125.
    (b) Determine the *minimum* effective depth needed if ρ can be changed.

7.2 A simply supported beam of 15-in width is 20 ft long and carries a uniform service dead load of 2 kips/ft and an applied concentrated service live load of 30 kips at midspan. Assuming an effective depth of 30 in and an overall depth of 34 in, determine the steel

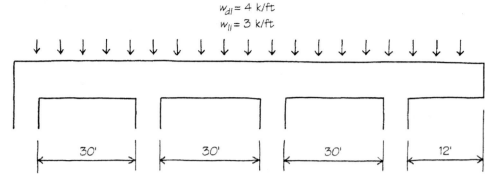

FIGURE 7.3 Beam for Problem 7.11.

ratio required at midspan. Let $f_y = 60$ ksi and $f'_c = 3000$ psi.

7.3 Determine the effective depth required for the cantilever beam of Example 7.5 if a lower ρ value of, say, $\frac{1}{2}\rho_{max}$, were to be used. Assume that all other conditions are unchanged.

7.4 A continuous beam 16 in wide with supports 25 ft apart has a design load w_u of 7 kips/ft, which includes the self-weight; moment at the supports is estimated at $wL^2/11$. Letting $f'_c = 4000$ psi, $f_y = 60\ 000$ psi, and $\rho = 0.0130$, find the required effective depth for the beam.

7.5 Without changing the strength of the steel or concrete, how could the beam of Problem 7.2 be redesigned so that the overall depth does not exceed 30 in?

7.6 Determine the required effective depth of a floor girder that spans the 30 ft-square bays of a warehouse. The floor is a one-way pan system that contributes a service dead load of 100 psf. The entire floor also carries a service live load of 200 psf. The girder's end moments are estimated at $wL^2/11$. Let $\rho = 3$ $\frac{2}{3}\rho$max, $f_y = 60$ ksi, $f'_c = 4000$ psi; and $b = 16$ in.

7.7 Find the approximate required effective depth for a continuous beam of 14-in width over 20-ft spans with uniform service loads of 3 kips/ft (live) and 4 kips/ft (dead; in-

cludes estimated beam self-weight). Maximum moment has been estimated to be $wL^2/9$. Assume that $\rho = \frac{2}{3}\rho_{max}$, $f_y = 60$ ksi, and $f'_c = 3$ ksi.

7.8 Find the approximate effective depth for the beam of Problem 7.7 if 5000-psi concrete is used instead of 3000-psi concrete.

7.9 A continuous beam of 35-ft span and 18-in width must support a uniform dead load of 2 kips/ft (not including self-weight), a concentrated dead load of 50 kips at midspan, and a live load of 4 kips/ft. The end moments due to the uniform loads may be taken as $wL^2/11$; those due to the point load may be taken as $PL/12$. Letting $f_y = 60\ 000$ psi, $f'_c = 5000$ psi, and $\rho = \frac{2}{3}\rho_{max}$, determine the approximate effective depth necessary to accommodate the maximum moment for this beam.

7.10 For preliminary design purposes, determine the approximate required effective depth for a simply supported rectangular reinforced concrete beam of 14-in width that carries a 15-kip dead load at the midpoint of its 25-ft span as well as a distributed live load of 5 kips/ft. Use $\rho = \frac{2}{3}\rho_{max}$, $f_y = 60$ ksi, and $f'_c = 4000$ psi.

7.11 Find the approximate effective depth that would be required for the continuous beam of Figure 7.3. The beam width is 17 in, $\rho = \frac{2}{3}\rho_{max}$, $f_y = 60$ ksi, and $f'_c = 3000$ psi. (*Hint:* First estimate the maximum moment.)

7.12 The data and beam of Problem 7.11 are unchanged except that the cantilevered span is now 14 ft long. Determine the approximate effective depth required after this change.

7.13 Select tensile reinforcement for the beam of Problem 7.1(*a*) and sketch the resulting section.

7.14 Select tensile steel for the beam of Example 7.4 and detail the reinforced section
(*a*) At one of the ends
(*b*) At midspan

8

SHEAR IN REINFORCED CONCRETE BEAMS

8.1 INTRODUCTION

This chapter is limited to an investigation of shear in beams and the placement of reinforcing to counteract diagonal tension stresses. The principles involved, however, also apply to shearing stresses found in the support brackets of precast construction and to the behavior of continuous shear walls used as lateral force resisting systems. (See Figure 8.1.) Punching shear, which occurs in column footings and thin floor slabs and results from concentrated loads tending to push through or puncture the concrete surface, is discussed in Chapter 13.

FIGURE 8.1

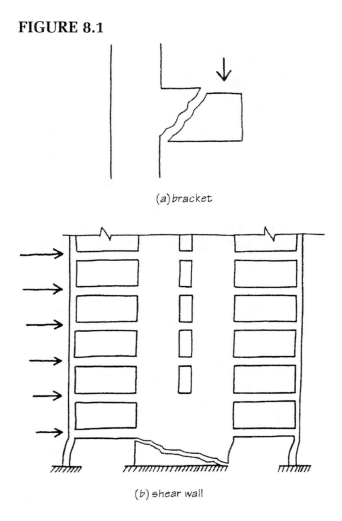

(a) bracket

(b) shear wall

8.2   DIAGONAL TENSION

The primary concern with shear in concrete beams is seldom the shearing stresses themselves, but rather the *diagonal tension*, which always accompanies such stresses. (See Figure 8.2.) From basic mechanics we know that vertical shear is always accompanied by an equal amount of horizontal shear (and vice versa) and that these pairs of shear forces form resultants which create diagonal tension. Concrete is capable of taking some tension but the typical beam requires special reinforcing to control these diagonal tension stresses. Figure 8.3 shows the crack pattern in a typical uniformly loaded simple beam in an overload condition. The cracks near the middle of the span tend to be more or less vertical because the shear forces there are very small and the moment forces are very large; there is little or no diagonal tension but a lot of horizontal tension due to the internal moment couple. On the other hand, the cracks become more sloping near the ends of the beam where the shear forces are larger and the moment forces become small. The crack at the extreme left end of the beam represents a common type of

FIGURE 8.2 Development of diagonal tension.

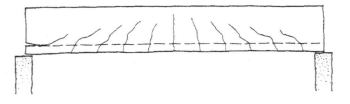

FIGURE 8.3 Typical crack pattern.

shear failure called "shear and bond splitting." The horizontal portion of the crack is caused by vertical shear forces acting across the flexural steel, tending to push it down and out of the beam. Figure 8.4 shows these cracks in a laboratory test beam.

8.3   STIRRUPS

The special reinforcing referred to earlier is usually in the form of U-shaped pieces of reinforcing bar called stirrups. These are almost always fabricated from #3 or #4 bar, and some typical shapes are shown in Figure 8.5. The two shapes in Figure 8.5*a* and *b* are the most common.

Typically, the spacing of stirrups will vary over the length of a beam depending upon the magnitude of the shear forces. They are sometimes placed on a diagonal as obviously it would be most effective to have them cross a potential crack at 90° the way moment steel does. However, for ease of construction stirrups are usually placed vertically and the spacing varies incrementally rather than evenly over the beam length. A typical pattern is shown in Figure 8.6. The designation for this spacing on a working drawing might be 1/2, 5/4, 6/8, 4/14, EE, meaning the first stirrup is 2 in from the end, followed by five stirrups spaced 4 in apart, followed by six stirrups 8 in apart, and ending with four spaced at 14 in apart. *EE* refers to "each end" of the beam and

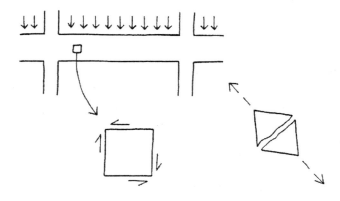

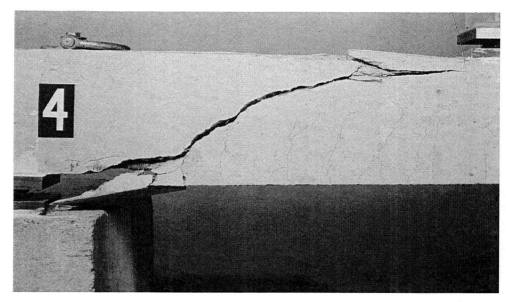

FIGURE 8.4 Explosive shear failure of a laboratory test beam. (*Professor William L. Gamble, Department of Civil Engineering, University of Illinois at Urbana-Champaign*)

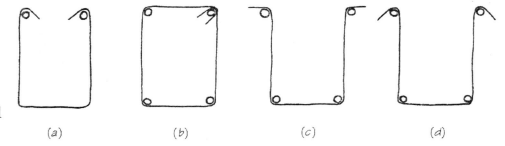

FIGURE 8.5 Typical stirrup shapes.

(a)   (b)   (c)   (d)

FIGURE 8.6 Typical stirrup placement.

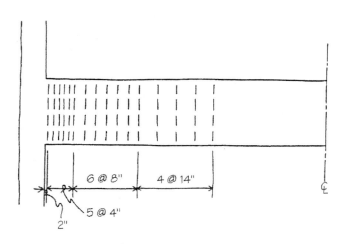

assumes a symmetrical loading pattern. The closest practical spacing is 3 or 4 in; and larger diameter stirrups or a larger beam cross section should be used if the magnitude of shear requires a closer spacing. It is customary practice to place the first stirrup at a half space, which accounts for the 2-in space in this illustration. The maximum spacing, 14 in in this case, is frequently controlled by a Code stipulation that states that if stirrups are needed in a beam, they cannot be spaced farther apart than $d/2$ (one-half the effective depth), or 24 in, whichever is smaller. This provision ensures that a potential diagonal tension crack (at any reasonable slope) will have to cross at least one stirrup as it forms. Additional stirrups are required whenever the longitudinal steel is terminated or reduced. In certain cases involving very large shear forces, maximum stirrup spacing is reduced to $d/4$ or 12 in, whichever is smaller.

8.4 DESIGN FOR SHEAR

The basic premise in shear design is that for any given section along the length of a beam the shear force V_u present due to the *factored* loads can be no greater than the capacity of the beam at that section. (As usual, the word "capacity" means the theoretical strength after reduction by the appropriate (ϕ factor.) This can be stated as

$$V_u \leq \phi(V_c + V_s) \qquad (8\text{-}1)$$

where V_c = shear capacity due to concrete
$\quad V_s$ = shear capacity due to the stirrups, if any
$\quad \phi$ = 0.75

The Code makes two exceptions, one when the shear forces are large and the other when they are small. Experiments show that when a beam is supported from below or constructed monolithically with a column in compression, localized compressive stresses that consequently exist in the beam near the support counteract the diagonal tension stresses. To account for this, the Code states that such beam sections do not have to be designed for the peak shear forces that usually exist at those places, but instead may be designed for the smaller shear forces present at a distance d (effective depth) from the support. The shaded areas of Figure 8.7 represent zones where the beam should be designed for the reduced shear forces "d distance out from the support" and not for the larger forces adjacent to the supports. (This reduction may *not* be taken if there are concentrated loads in those zones.) The cross section at d distance from the face of the support is often called the *critical section*.

Equation (8-1) is also not entirely applica-

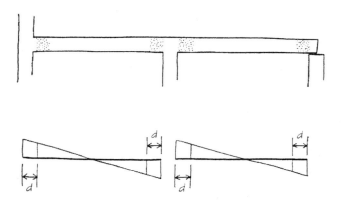

FIGURE 8.7

ble where the shear is small, e.g., near midspan of most beams carrying uniform loads and near the ends of cantilever beams carrying similar loads. In such zones Equation (8-1) indicates that when V_u becomes less than or equal to ϕV_c, no stirrups would be needed. To be conservative, since shear failures tend to be very rapid and without warning and since the strength of concrete can vary within a given beam, the Code requires that stirrup reinforcing be continued until V_u drops below *one-half* of ϕV_c.

The precise contribution of the concrete to the shear capacity (V_c) of a beam can sometimes be complicated to determine; fortunately the ACI Code permits a simple conservative expression to be used in most cases:

$$V_c = 2\sqrt{f'_c}\,bd \qquad (8\text{-}2)$$

where b = beam width (in)
$\quad d$ = effective depth (in)
$\quad f'_c$ = cylinder strength (psi)

(It is important that the units of f'_c be expressed in psi and not ksi whenever the square root is taken.)

The contribution of the stirrups can be more precisely determined, being merely a function of the number and size of stirrups and the strength of the steel. Figure 8.8 shows

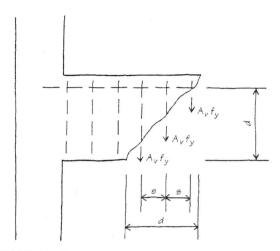

FIGURE 8.8

a diagonal tension crack at an assumed angle of 45° crossing several stirrups. If the stirrup spacing is given as s, then the number of stirrups crossed by the crack is d/s. The total contribution of the stirrups is then the strength of each times the number of stirrups, or

$$V_s = \frac{A_v f_y d}{s} \qquad (8\text{-}3)$$

where A_v = cross-sectional area of one stirrup (equal to two bar areas, assuming each stirrup has two legs)

f_y = yield strength of the steel
d = effective depth (in)
s = spacing of stirrups (in)

Assuming a constant beam section, V_c will be constant over the beam span while V_s will vary to reflect the changes in V_u. For ease of construction the size of the stirrups will usually be constant for a given beam so that the spacing varies as discussed in Section 8.3.

Equation (8-1) can be solved for V_s to get

$$V_s = \frac{V_u}{\phi} - V_c \qquad (8\text{-}1a)$$

and Equation (8-3) can be transposed to get

$$s = \frac{A_v f_y d}{V_s} \qquad (8\text{-}3a)$$

and the required spacing for any value of V_u is easily determined.

As previously mentioned, the spacing is subject to an upper limit of $d/2$. In order to prevent immediate (and unsafe) yielding of the stirrups when a diagonal crack forms, the Code also controls the maximum spacing of stirrups as a means of specifying a minimum amount of steel area. As the concrete strength increases, the fraction of the shear capacity provided by the concrete increases and a less ductile (more sudden) failure is likely. Therefore, the maximum spacing is controlled by equations (8-4).

$$S_{max} = \text{the smaller of} \begin{cases} \dfrac{d}{2} & (8\text{-}4a) \\[2ex] \dfrac{A_v f_y}{50b} & (8\text{-}4b) \\[2ex] \dfrac{A_v f_y}{0.75\sqrt{f_c'}\,b} & (8\text{-}4c) \end{cases}$$

where f_y and f_c' are both in psi.

Figure 8.9 shows the different areas for stirrup spacing requirements on the V diagram of a uniformly loaded beam.

The following examples will serve to illustrate stirrup placement. The decision as to where and how often to change the spacing of stirrups is arbitrary as long as the capacity provided is greater than that required. However, changing spacing frequently (to follow the shear diagram closely) can make field construction more difficult and will only result in significant savings if a large number of identical beams is involved on the project.

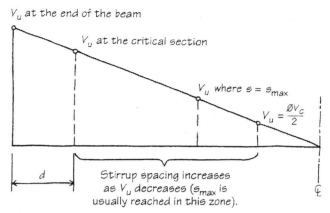

FIGURE 8.9

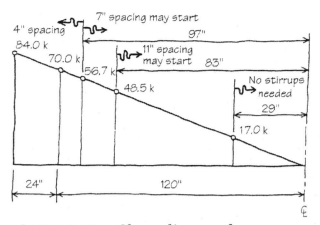

FIGURE 8.10   Shear diagram for Example 8.1

To simplify the process, the examples and problems of this chapter ignore the effects of member continuity upon the design shear values. This consideration is, however, addressed in the design of the beams and girders in Chapter 16.

Example 8.1   A 24-ft-long continuous beam must carry a uniform load of $w_u = 7$ kips/ft, which includes its own self-weight. Locate #3 stirrups if $f'_c = 3500$ psi, $f_y = 40$ ksi, $b = 16$ in, and $d = 24$ in.

Solution: The maximum V_u value will be 84.0 kips. Using the shear diagram of Figure 8.10, first locate the critical section and then locate the place where stirrups are no longer needed. The critical section will be 24 in in from the left end and the accompanying V_u value will be

$$\frac{120}{144}(84.0) = 70 \text{ kips}$$

Stirrups can stop where $V_u = \phi V_c /2$:

$$V_c = 2\sqrt{f'_c}bd$$
$$= 2\sqrt{3500}(16)(24)$$
$$= 45\,400 \text{ lb}$$

$$V_u = \frac{\phi V_c}{2}$$

$$= \frac{0.75(45.4)}{2}$$

$$= 17.0 \text{ kips}$$

Using similar triangles, we get

$$\frac{x}{17.0} = \frac{144}{84.0}$$

$$x = 29.1 \quad \text{(rounding down)} \approx 29 \text{ in}$$

Thus, we see that stirrups will not be needed in the central 58 in of this beam.

The maximum spacing permitted will be

$$\frac{d}{2} = \frac{24}{2} = 12 \text{ in} \qquad \text{(8-4a)}$$

or $\quad \dfrac{A_v f_y}{50b} = \dfrac{0.22(40\,000)}{50(16)} = 11 \text{ in}$ (8-4b)

or $\quad \dfrac{A_v f_y}{0.75\sqrt{f'_c}b} = \dfrac{0.22(40\,000)}{0.75\sqrt{3500}(16)} = 12.4 \text{ in}$

(A_v may be obtained from Table A.1 in the Appendix.)

So, $s_{\max} = 11$ in.

Now determine where the 11-in spacing can start. All spacings from this location back to the support will be smaller. The 11-in spacing will give us

$$V_s = \frac{A_v f_y d}{s} \qquad (8\text{-}3)$$

$$= \frac{0.22(40)(24)}{11}$$

$$= 19.2 \text{ kips}$$

This will enable us to find

$$V_u = \phi(V_c + V_s) \qquad (8\text{-}1)$$

$$= 0.75(45.4 + 19.2)$$

$$= 48.5 \text{ kips}$$

Again using similar triangles, we obtain

$$\frac{x}{48.5} = \frac{144}{84.0}$$

$$x = 83.1 \quad (\text{rounding down}) \approx 83 \text{ in}$$

Since the beam half is 144 in long, this point will be 61 in from the support.

Now find the *minimum* required spacing (at the critical section) where $V_u = 70$ kips. The capacity needed from the stirrups will be

$$V_s = \frac{V_u}{\phi} - V_c \qquad (8\text{-}1a)$$

$$= \frac{70.0}{0.75} - 45.4$$

$$= 47.9 \text{ kips}$$

Then we can obtain s:

$$s = \frac{A_v f_y d}{V_s} \qquad (8\text{-}3a)$$

$$= \frac{0.22(40)(24)}{47.9}$$

$$= 4.4 \text{ in} \quad (\text{rounding down}) \approx 4 \text{ in}$$

Where practical, the next smaller spacing to the *nearest inch* should be used. The first stirrup should be placed a distance of $s/2$ from the face of the support. To keep from having to bother with ½-in increments, we shall place the first stirrup at 2 in and then carry the 4-in spacing out to where an intermediate spacing can take over. Logical choices for this intermediate spacing (between the 4- and 11-in spacings) will be 7 or 8 in. Trying 7 in, find out where this spacing can begin exactly the way we did for the 11-in maximum spacing:

$$V_s = \frac{A_v f_y d}{s}$$

$$= \frac{0.22(40)(24)}{7}$$

$$= 30.2 \text{ kips}$$

Then

$$V_c = \phi(V_c + V_s)$$

$$= 0.75(45.4 + 30.2)$$

$$= 56.7 \text{ kips}$$

$$\frac{x}{56.7} = \frac{144}{84.0}$$

$$x = 97 \text{ in}$$

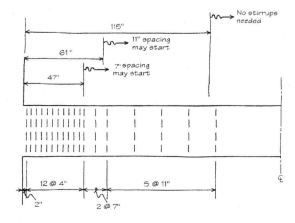

FIGURE 8.11 Stirrup spacing for Example 8.1

Again, the beam half is 144 in long, so this point will be 47 in from the support. We will need to carry the 4-in spacing at least to 47 in, then carry the 7-in spacing at least to 61 in (144 – 83), where the 11-in spacing can start as previously computed. Figure 8.11 illustrates the selected spacings and distances. An appropriate designation would be 1/2, 12/4, 2/7, 5/11 EE.

Example 8.2   Determine the spacing pattern of #4 stirrups for the continuous girder in Figure 8.12. The loads shown are service loads and include all self-weight; f'_c = 4000 psi, f_y = 40 ksi, b = 14 in, and d = 28 in.

Solution: First factor the loads and establish the shear diagram of Figure 8.13:

$$w_u = 1.2(1.6) = 1.92 \text{ kips/ft}$$

$$P_u = 1.2(20) + 1.6(14) = 46.4 \text{ kips}$$

$$V_{max} = 13.5(1.92) + 46.4 = 72.3 \text{ kips}$$

The V_u value at the critical section will be obtained by knowing that the slope of the diagram is 1.92 kips/ft. In 2.33 ft, the V value will drop by an amount equal to 2.33 ft (1.92 kips/ft) = 4.5 kips. The V_u value at the critical section will be 67.8 kips.

Stirrups can stop where $V_u = \phi V_c/2$:

$$V_c = 2\sqrt{f'_c}bd$$
$$= 2\sqrt{4000}\,(14)(28)$$
$$= 49\,600 \text{ lb}$$

$$V_u = \frac{\phi V_c}{2}$$
$$= \frac{0.75(49.6)}{2}$$
$$= 18.6 \text{ kips}$$

This value occurs at the point load where the V diagram drops from 55.0 to 10.0 kips, and thus no stirrups will be needed in the middle third of the girder.

The maximum spacing permitted will be

$$\frac{d}{2} = \frac{28}{2} = 14 \text{ in}$$

or $$\frac{A_v f_y}{50b} = \frac{0.39(40\,000)}{50(14)} = 22.3 \text{ in}$$

or $$\frac{A_v f_y}{0.75\sqrt{f'_c}b} = \frac{0.39(40\,000)}{0.75\sqrt{4000}(14)} = 23.5 \text{ in}$$

So, s_{max} = 14 in.

FIGURE 8.12

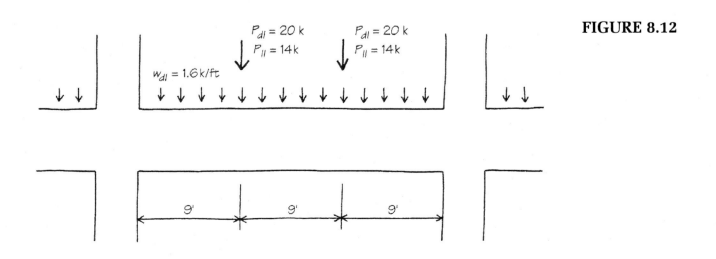

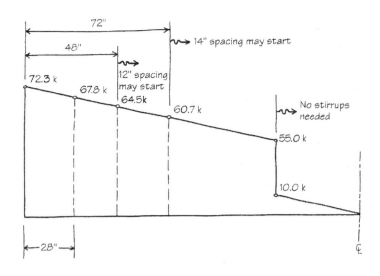

FIGURE 8.13 Shear diagram for Example 8.2

To determine where this 14-in spacing can start, we calculate

$$V_s = \frac{A_v f_y d}{s}$$

$$= \frac{0.39(40)(28)}{14}$$

$$= 31.2 \text{ kips}$$

$$V_u = \phi(V_c + V_s)$$

$$= 0.75(49.6 + 31.2)$$

$$= 60.7 \text{ kips}$$

FIGURE 8.14 Stirrup spacing for Example 8.2

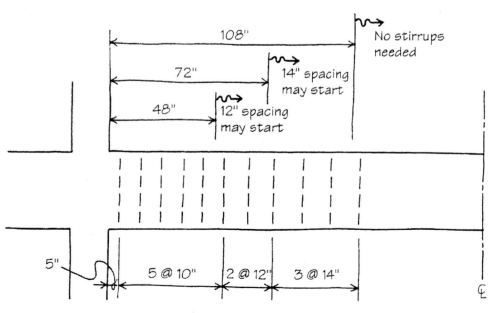

To locate the place where the ordinate has this value on the shear diagram, we note that 60.7 is 11.6 less than the end value of 72.3 kips. At a slope of 1.92 kips/ft it will take

$$\frac{11.6}{1.92} = 6.04 \, \text{ft} \quad \text{or} \, 72 \, \text{in}$$

for the diagram to drop to 60.7 kips. Between this section and the point load we can use a 14-in spacing.

At the critical section we can find the required spacing by getting

$$V_s = \frac{V_u}{\phi} - V_c$$

$$= \frac{67.8}{0.75} - 49.6$$

$$= 40.8 \, \text{kips}$$

$$s = \frac{A_V f_y d}{V_s}$$

$$= \frac{0.39(40)(28)}{40.8}$$

$$= 10.7 \, \text{in} \quad (\text{rounding down}) \approx 10 \, \text{in}$$

Let's see if the spacing can change to 12 in as we move away from the support:

$$V_s = \frac{A_v f_y d}{s}$$

$$= \frac{0.39(40)(28)}{12}$$

$$= 36.4 \, \text{kips}$$

$$V_u = \phi(V_c + V_s)$$

$$= 0.75(49.6 + 36.4)$$

$$= 64.5 \, \text{kips}$$

This value is 7.8 kips less than 72.3 kips, so its location will be

$$\frac{7.8}{1.92} = 4.06 \, \text{ft} \quad \text{or} \, 48 \, \text{in}$$

from the end.

Placing the first stirrup at 5 in from the end, we carry the 10-in spacing at least out to 48 in, then change to 12-in spacing, carrying it past 72 in, where the spacing will become 14 in (Figure 8.14). An appropriate designation would be 1/5, 5/10, 2/12, 3/14 EE.

8.5   TORSION

Torsional forces acting on beams also produce shearing stresses. Small torsional forces are usually present in almost every bending member due to the improbability of all loads being applied in a symmetrical pattern. Larger torsional forces exist when members must be deliberately loaded in an eccentric fashion. The most readily observable example of eccentric loading occurs in every spandrel beam. Under full design load the interior beams of the section shown in Figure 8.15 are symmetrically loaded. The spandrel beam, however, is loaded only from one side and is therefore subjected to torsional as well as bending forces.

The resulting shearing stresses produce diagonal tension stresses which require additional reinforcement. Unlike flexural shear, which becomes zero at the top and bottom edges of a beam, the stresses due to torsion act on all surfaces of the cross section. Thus, whenever torsional stresses require the use of stirrups, a fully closed configuration must be used, such as that shown in Figure 8.5*b*. It is

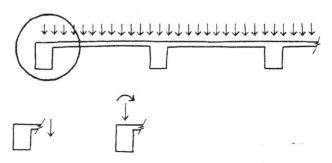

FIGURE 8.15

also desirable to avoid narrow deep cross sections, as these are less efficient in torsion than those that are more square.

A beam which develops torsional cracks tends to increase in length as the cracks open up. For this reason stirrups placed to counteract torsional stresses are always accompanied by longitudinal steel. In addition to the bars normally present as moment steel, the Code requires that longitudinal bars be placed at intervals not to exceed 12 in on all sides of the member.

The design procedures for the proper placement of steel for torsion involve a consideration of the interactive effects of shear and torsion in the member and are somewhat beyond the scope of this basic text. For the purposes of structural planning and preliminary design, however, it should be noted that torsional forces seldom control the overall size of the member. That is, if the beam or girder is properly sized to carry the bending moments and shears and is not overly narrow, then the torsional forces can usually be handled by the use of additional reinforcing.

8.6   STIRRUPS AS TIES

In Chapter 7 the use of longitudinal bars on the compression side of flexural members was discussed. Such bars can serve to reduce the required depth of a member and to combat long-term deflection due to creep. Just as in a column when long slender bars are used in compression, they must be prevented from buckling outward. In a beam the stirrups serve the valuable function of restraining the compression steel to prevent such a failure.

PROBLEMS

8.1 Using the beam of Example 8.1, determine the spacing of stirrups if #4 reinforcing bar is used.

8.2 A reinforced concrete simple beam supports a total factored uniform load of 5 kips/ft over its 40-ft span. Determine a reasonable stirrup pattern using #4 bar. Beam width is 16 in, effective depth is 30 in, f'_c = 4500 psi, and f_y = 40 000 psi.

8.3 For a simply supported rectangular beam of 20-ft span, carrying a uniform service dead load of 4 kips/ft (does not include s.w.), what beam width would be required if no stirrups were used for shear reinforcement? The effective depth of the beam is 18 in and f'_c = 5000 psi. Ignore shear reduction.

8.4 The 8-ft cantilever of Figure 8.16 carries a uniform load of 0.5 kip/ft and a concentrated load of 40 kips at midpoint. Both loads are service dead loads and the self-weight of the beam is included. Beam width is 10 in, d = 20 in, f'_c = 4500 psi, and f_y = 60 000 psi. Determine stirrup placement using #3 bar.

FIGURE 8.16

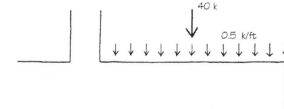
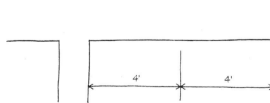

8.5 Determine a reasonable spacing pattern using #3 stirrups for the beam shown in Figure 8.17. The loads are service live loads and do not include self-weight; $b = 12$ in, $d = 24$ in, $f_y = 6.0$ ksi, and $f'_c = 4500$ psi.

8.6 A roof girder on a 30-ft simple span carries a uniform ultimate design load of 3 kips/ft (includes weight of beam) and three 32-kip point loads: one at midspan, the other two at 2 ft from each end. Find a reasonable spacing pattern for #3 stirrups if $f_y = 40$ ksi, $f'_c = 4000$ psi, $b = 16$ in, and $d = 32$ in.

FIGURE 8.17 Beam loading for Problem 8.5.

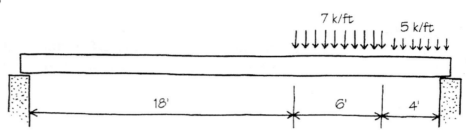

9
DEFLECTION

9.1 INTRODUCTION

Relative to timber and steel, reinforced concrete structural elements tend to be quite stiff. The modulus of elasticity value is two or three times that of wood, and compared to steel, concrete sections embody much larger moments of inertia. The monolithic nature of construction also contributes to reduced deflections as the negative end moments in continuous beams serve to counteract the positive bending at midspan.

Nevertheless, shallow members can have significant deflection problems. The ACI Code provides Table 11.1 of minimum thicknesses. When the overall depth is less than the ratio given *or* when deflections cause cracking of ceilings or damage to attached partitions or doors and windows, deflections must be calculated to make sure they do not exceed certain limits. Such limits are often given as a fraction of the span such as $L/240$ or $L/360$.

In concrete there are two kinds of deflection, the immediate or instantaneous deformation that takes place upon application of the load and the long-term deformation associated with shrinkage and *creep*. Creep is time dependent and is caused only by the *sustained* loads; usually this means all the dead load plus a small fraction of the live load. (However, in the case of a warehouse, for example, the sustained load might include most of the live load.) Creep is directly related to the amount of compression stress in the concrete, and the best way to combat creep is through the use of steel in the compression zone of a beam to reduce those stresses. Such steel is also effective in controlling deflection due to shrinkage.

The loads referred to earlier are, of course, the *service* loads and not the factored loads needed for strength design. Excess deflection is a *serviceability* problem rather than a strength problem.

9.2 DEFLECTION COMPUTATIONS

Nonstructural components such as ceilings, partitions, doors, and windows are usually attached to the structure after a significant portion of the sustained load has been applied. The Code limits on deflection recognize this, and one of the more stringent limits ($L/480$) applies only to deflections occurring after these nonstructural elements are in place; that would be the long-term deflection due to the sustained loads plus the immediate deflection due to the remainder of the live load. To simplify computations it will usually suffice to sum the long-term dead load deflections and the immediate live load deflection.

The biggest hurdle in finding such deflections is the determination of an appropriate moment of inertia to use. As shown in Figure 9.1, a typical beam under service load conditions has numerous small cracks whenever the moment is large. Obviously, the gross moment of inertia for a rectangle, computed as $I_g = bh^3/12$, would apply only to zones of low moment where there were no cracks. Its use for the entire beam would result in very small deflections. Where the beam is cracked, we could use the so-called *cracked section* moment of inertia, I_{cr}, which assumes that once a crack has occurred, it propagates immediately to the neutral axis, i.e., to the base of the compressive stress zone.

The Code recommends the use of an effective moment of inertia, I_e, which has a value between the cracked and the gross values and is computed using an empirical formula. The effective moment of inertia is dependent upon I_{cr} and I_g and will vary over the beam length, so the Code recommends that an average of the end and midspan sections be used. The numerical manipulations can be considerable and a number of design aids and computer programs has been developed to assist the beam designer.

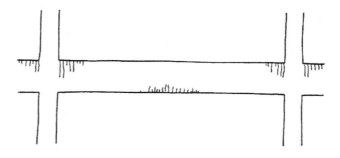

FIGURE 9.1 Representative moment cracks under service loads.

For preliminary deflection checks we find that the use of the midspan cracked section I value will provide sufficient accuracy and will err on the conservative side. (The value at the supported end of a cantilever should be used.) This approach has been used in the examples of this chapter.

With the I value ascertained (more or less), the use of standard deflection formulas (such as those found in Table D of the Appendix) will be adequate to determine the immediate deflection in most situations. As always it is up to the user to exercise judgment as to when these formulas are directly applicable and when they are providing values that are inconsistent with the actual loading and support conditions.

The long-term or creep deflection can be found by multiplying the immediate deflection by a factor dependent upon the amount of compression steel present:

$$\Delta_{lt} = \frac{2}{1 + 50\rho'}(\Delta_i) \qquad (9\text{-}1)$$

where $\rho' = \dfrac{A'_s}{bd}$, in which A'_s is the compression steel area

Example 9.1 The beam of Figure 9.2 is attached to partitions which may be damaged if the deflec-

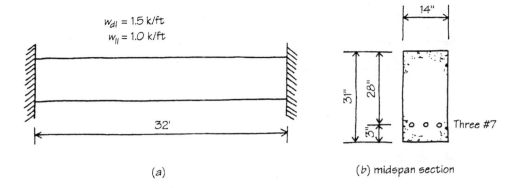

FIGURE 9.2   Beam for Example 9.1.

tion exceeds $L/480$. Ascertain whether or not the beam will meet this criterion by using the midspan I_{cr}. Let $f'_c = 3000$ psi.

Solution: Add the long-term deflection due to the dead load to the immediate deflection for the live load and compare the sum to $L/480$. First obtain the cracked section moment of inertia by getting an equivalent all concrete section. This fictitious all concrete section is called the *transformed section* and is homogeneous, making it possible to compute an I value. The steel can be transformed into an equivalent amount of concrete by using a ratio of the two moduli of elasticity values. As explained in Chapter 2, the E value for regular stone concrete is approximately

$$E_c = 57\,000\sqrt{f'_c} \qquad (2\text{-}2)$$

$$= 57\,000\sqrt{3000}$$

$$= 3.1(10)^6 \text{ psi}$$

The E value for steel is, of course, $29(10)^6$ psi, so the modular ratio is

$$n = \frac{E_s}{E_c}$$

$$= \frac{29(10)^6}{3.1(10)^6}$$

$$= 9 \quad \text{(rounded to the nearest whole number)}$$

The three #7 bars have an area of 1.8 in^2, so the equivalent area of concrete will be

$$A_c = nA_s$$

$$= 9(1.8)$$

$$= 16.2 \text{ in}^2$$

The idea behind this transformation is that since the two materials are bonded together, they must undergo the same strain. If they have equal strains, they will develop stresses in direct proportion to their modulus of elasticity values. If the new equivalent area of concrete is to provide the same resistance in terms of force as the replaced steel, it will need to have an area n times as large. (The new concrete is placed at the same distance from the neutral axis as the replaced steel, so it will indeed have the same strain.) Figure 9.3*b* shows the transformed section for this example. The values were obtained by solving a quadratic for the location of the neutral axis:

$$\frac{b\bar{y}^2}{2} = nA_s(d - y)$$

$$\frac{b\bar{y}^2}{2} + nA_s y - nA_s d = 0$$

$$7\bar{y}^2 + 16.2y - 454 = 0$$

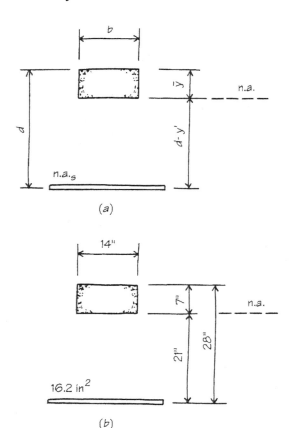

FIGURE 9.3 Transformed section.

$$\overline{y} = \frac{-b \pm \sqrt{b^2 - 4ac}}{2a}$$

$$= \frac{-16.2 \pm \sqrt{(16.2)^2 - 4(7)(-454)}}{2(7)}$$

$$= 7 \text{ in}$$

The value of I_{cr} will then be found by using the base moment of inertia term for the compression block plus the parallel axis theorem for the transformed steel. (The centroidal term for the latter will be omitted as negligible.)

$$I_{cr} = \frac{b\overline{y}^3}{3} + nA_s(d - \overline{y})^2$$

$$= \frac{14(7)^3}{3} + 16.2(21)^2$$

$$= 8740 \text{ in}^4$$

From Table D of the Appendix, the formula for the deflection of a uniformly loaded fixed end beam is

$$\Delta = \frac{wL^4}{384 \, EI}$$

The immediate deflection due to the live load is

$$\Delta_{i_{ll}} = \frac{1.0(32)^4(12)^3(1000)}{384(3.1)(10)^6(8740)}$$

$$= 0.17 \text{ in}$$

The immediate deflection due to the dead load is proportional

$$\Delta_{i_{ll}} = \frac{1.5}{1.0}(0.17) = 0.26 \text{ in}$$

Since there is no compression steel, the factor from Equation (9-1) will be 2.0. The total deflection that may contribute to damaging the partitions will be

$$\Delta_t = 2.0(0.26) + 0.17$$

$$= 0.69 \text{ in}$$

The permissible deflection by the Code is $L/480$:

$$\frac{L}{480} = \frac{32(12)}{480} = 0.80 \text{ in}$$

Since 0.69 is less than 0.80, the beam as reinforced will be okay in terms of deflection.

Example 9.2 The cantilever beam of Example 7.6 carries library stacks and it is estimated that 75% of the live load will occur as a sustained load on the beam. (See Figure 9.4.) Four #9 bars were added to the bottom of this beam to help combat long-term deflection. Compute the probable maximum deflection of this beam.

Solution: The f_c' value from Example 7.6 is 4000 psi. The E value for the concrete is then estimated by Equation (2-2) to be

$$E_c = 57 \, 000\sqrt{4000}$$

$$= 3.6(10)^6 \text{ psi}$$

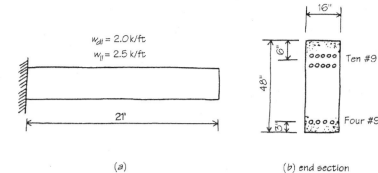

FIGURE 9.4 Beam for
Example 9.2.

(a)

(b) end section

which gives us a modular ratio of

$$n = \frac{E_s}{E_c}$$

$$= \frac{29(10)^6}{3.6(10)^6}$$

$$= 8$$

The transformed end section appears in Figure 9.5. The compression steel was transformed using a multiplier of $n - 1$ rather than n because the holes vacated by the reinforcing bars are now replaced by part of the concrete in the main compression block:

Tension steel:

$$A_c = nA_s$$

$$= 8(10)$$

$$= 80 \text{ in}^2$$

Compression steel:

$$A'_c = (n - 1)(A'_s)$$

$$= 7(4)$$

$$= 28 \text{ in}^2$$

FIGURE 9.5   Transformed section.

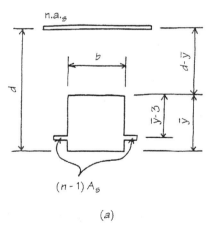

(a)

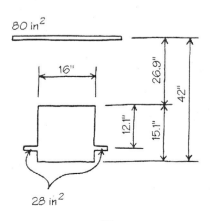

(b)

To find the $\overline{y}$ value (with respect to the bottom edge in this case), we calculate

$$\frac{b\overline{y}^2}{2} + (n-1)A'_s(\overline{y} - 3) = nA_s(d - \overline{y})$$

$$8\overline{y}^2 + 28\overline{y} - 84 = 3360 - 80\overline{y}$$

$$8\overline{y}^2 + 108\overline{y} - 3444 = 0$$

$$\overline{y} = \frac{-108 \pm \sqrt{(108)^2 - 4(8)(-3444)}}{2(8)}$$

$$= 15.1 \text{ in}$$

We can then find I_{cr}:

$$I_{cr} = \frac{16(15.1)^3}{3} + 28(12.1)^2 + 80(26.9)^2$$

$$= 80\,300 \text{ in}^4$$

The formula for the deflection of a uniformly loaded cantilever is

$$\Delta = \frac{wL^4}{8EI}$$

The immediate deflection due to the sustained load (the dead load plus 75% of the live load) is

$$\Delta_{i_{sl}} = \frac{(2.0 + 1.9)(21)^4(12)^3(1000)}{8(3.6)(10)^6(80\,300)}$$

$$= 0.57 \text{ in}$$

The immediate deflection due to the remaining live load is by ratio

$$\Delta_{rem_{ll}} = \frac{0.6}{3.9}(0.57)$$

$$= 0.09 \text{ in}$$

The total deflection will be composed of three parts: the immediate deflection from the sustained load, the long-term effect of the sustained load, and the immediate deflection of the remaining live load. To get the second of these components, we need to find the ρ' value for the compression steel:

$$\rho' = \frac{A'_s}{bd}$$

$$= \frac{4}{16(45)}$$

$$= 0.0055$$

$$\Delta_{lt} = \frac{2}{1 + 50\rho'}(\Delta_i)$$

The total deflection then will be

$$\Delta_t = \Delta_{i_{sl}} + \Delta_{lt} + \Delta_{rem_{ll}}$$

$$= 0.57 + \left[\frac{2}{1 + 50(0.0055)}\right]0.57 + 0.09$$

$$= 1.55 \text{ in}$$

This deflection would be in excess of any of the Code limits (even though most of the first 0.57 in will take place before the nonstructural parts are attached) and the design would have to be modified, most likely by increasing both b and d of the cross section. The cantilever is probably too long in the first place and there will be a severe headroom problem in any event.

PROBLEMS

9.1 The only applied load that a cantilever beam 15 ft long carries is a concentrated dead load of 5 kips at its free end. The beam section is 10 × 19 in with an effective depth of 16 in; it is reinforced for tension with three #8 bars in one row. Letting $f_y = 60$ ksi and $f'_c = 3500$ psi, calculate
  (a) The immediate deflection of the cantilever
  (b) The total long-term deflection of the cantilever

9.2 A simply supported beam 40 ft long supports a uniform live load of 1.5 kips/ft and a concentrated dead load of 15 kips at midspan. The beam is 20 in wide, has an effective depth of 36 in and an overall depth of 40 in, and is reinforced for tension with eight #9 bars (60 ksi steel) in two rows. f'_c is 5000 psi. Will this beam stay within a specified initial deflection limit of $L/360$?

9.3 A girder 16 × 34 in in cross section has fixed ends and must support a dead load of 3 kip/ft over its 30-ft span. It is reinforced with five #8 bars at midspan and has an effective depth of 31 in. If 80% of its live load must be considered as a sustained load, how much uniform live load can the beam support while remaining within a permissible total deflection limit of $L/360$? Let $f'_c = 4000$ psi.

9.4 Will the total deflection of a simply supported beam reinforced as shown in Figure 9.6 stay within the specified limit of $L/240$? The beam is 40 ft long; it carries a live load of 0.5 kip/ft and a dead load of 1.0 kip/ft. Let $f'_c = 5500$ psi.

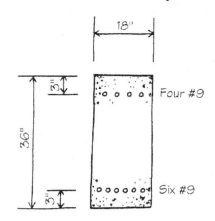

FIGURE 9.6

10

COLUMN BEHAVIOR

10.1 BEAM-COLUMN ACTION

Most building columns regardless of the structural material act as *beam-columns*. That is, they must sustain the effects of bending as well as axial forces. The loads that come into a column from the column above are generally axial, but loads that are brought in from the beam and girders attached to the sides of a column are eccentric to the axis of a column. Figure 10.1 illustrates how an eccentric load results in a couple applied to the column. Depending upon the magnitude of the eccentricity, the equivalent couple may be small or large.

If the connection is a moment-resistant one involving continuity (seldom in wood, sometimes in steel, but almost always in poured-in-place concrete), then the columns act together with the beams to share moments in the frame and the magnitude of the moment in a column is usually quite significant. (Exterior columns also serve as beams when the wind load acting via the building skin applies a uniform load along the length of the column, but that is not the action being discussed here.) Figure 10.2 shows how different loading patterns can result in different magnitudes and signs of end moments. Columns must be able to accommodate these changing moments and are designed for "worst case conditions" under

FIGURE 10.1

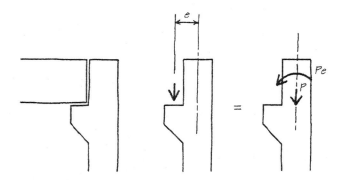

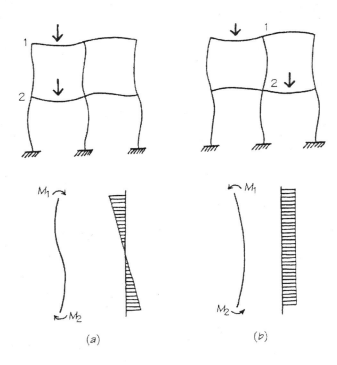

FIGURE 10.2 Influence of loading patterns upon column end moments.

several loading patterns. Interior columns usually have smaller end moments because the dead load would be present in all bays of a frame and tend to temper the moments shown in Figure 10.2*b*. Almost all exterior columns must be designed for the reverse curvature effect which results from being loaded from only one side.

In multistory construction lower-story columns have large axial forces compared to upper-level columns. Both have about the same amount of moment from the girders, so relatively speaking, the effect of the moment on column design gets greater as you go up in height. For simplicity of construction columns are often made with a constant cross section for the full height of the building (except in very tall structures) and the strength of the concrete and the amount of reinforcement is varied as required by the loads.

10.2 THE *P*-DELTA EFFECT

If the end moments result in significant deformation of the column midheight, or if the frame is unbraced such that it carries lateral loads via *sidesway*, then the effect of such displacements can influence the column design. Figure 10.3 illustrates two variations of the so-called *P-delta* effect in which the lateral displacement of a column results in an additional moment from the axial forces.

10.3 TYPES OF COLUMNS

The American Concrete Institute (ACI) Code classifies two basic types of columns by the way in which the longitudinal steel is tied to prevent it from buckling out of the column. These are called *laterally tied* and *spirally tied* and are illustrated in Figure 10.4. (They are sometimes simply referred to as "tied" and "spiral" columns, respectively.) Laterally tied columns are usually square or rectangular and spirally tied columns are often circular or octagonal, but either reinforcing pattern can be used with any outside envelope.

FIGURE 10.3 The *P*-delta effect.

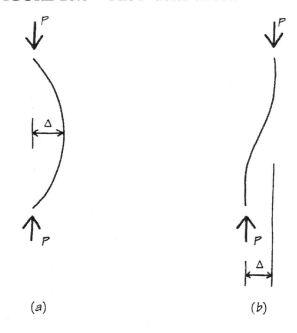

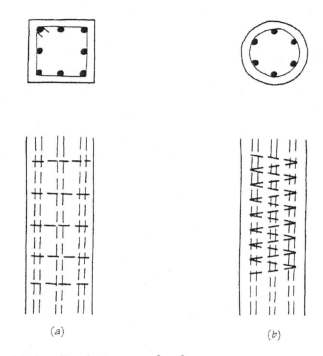

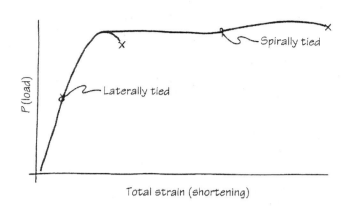

FIGURE 10.5 Load versus deformation.

FIGURE 10.6 Actual columns after a load test. (Note the greater shortening achievable with the spirally tied column.)

10.4 Basic types of columns.

Laterally tied columns are more frequent because they cost less, and there is little difference in the maximum load capacity of the two columns. Given the same amount of steel, the spirally tied column will be tougher or more *ductile*; i.e., it will deform more before it finally collapses. Figure 10.5 compares the approximate load versus axial deformation curves for the two types of columns. It is evident that the spirally tied column will provide a slower failure with the probability of providing some warning before it gives way, and the Code recognizes this by assigning a larger ϕ factor to spirally tied columns. The spiral pattern provides better containment for the inner core of concrete and the column will continue to support load after the outer "shell" has spalled off. The spiral column's increased ductility makes it capable of absorbing more energy and a spirally tied column will perform better under seismic loading conditions than a laterally tied column. Figure 10.6 shows the two types of columns after they were tested to failure.

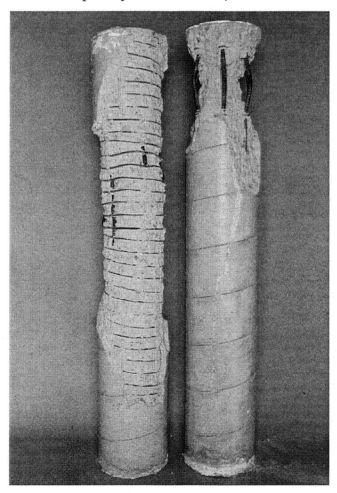

10.4 REINFORCING AND TIE REQUIREMENTS

The ρ value for columns is based upon the overall dimensions of the cross section and is bounded by ρ_{min} of 1% and ρ_{max} of 8%. Most columns are reinforced in the 2 to 6% range because of the difficulties associated with the placement of concrete in a form crowded with steel. The clear distance between bars must be at least as great as 1.5 times the bar diameter or $1\frac{1}{2}$ in, whichever is greater. For more heavily reinforced columns, the Code permits the use of "bundled" bars where two or three bars are wired side by side to allow more space for the placement of concrete. Some columns are made large for architectural reasons and in such cases, provided adequate strength can be demonstrated, the Code will permit the use of less than 1% steel. However, a minimum of four bars is required for a laterally tied column and a minimum of six for a spirally tied column. Table 10.1 provides some practical information regarding the "tie" steel. For rectangular columns with lateral ties the Code requires that every corner bar and every other intermediate bar be tied by an included angle no greater than 135°. No bar may be more than 6 in from a tied bar. Figure 10.7 shows some typical bar patterns. (Lateral ties may be in a circular pattern when surrounding a circle of longitudinal bars.)

10.5 PRELIMINARY DESIGN GUIDELINES

The basis for the design of any beam-column is an interaction curve. Figure 10.8 shows a representative curve for a reinforced concrete beam-column. In general, a member that has a lot of axial load can take only a small amount of moment, and a member with a large moment acting on it can accommodate only small axial loads. In the zone above the dashed line of Figure 10.8 compression controls and the member would fail primarily by crushing. In the area below the dashed line tension controls and the member would fail more like a beam than a column. This lower region of the curve is very interesting in that it indicates that a member carrying a large moment will actually be able to take more moment if some axial load is applied. (This is, of course, the basis of all prestressing operations in beams.)

Typical design handbooks have many such interaction graphs for different conditions; usually each graph is a series of curves moving outward from the origin, each one representing a larger cross section or an increased steel percentage. Computer programs to assist in designing columns make use of data used to plot such curves. A typical set of curves would be like that shown in Figure 10.9. P_r and M_r are the axial load and moment capaci-

TABLE 10.1 *Tie Reinforcement Provisions*

	Lateral	**Spiral**
Size of tie	# 3 or # 4 depending upon the size of the longitudinal bars	Determined by formula, usually $\frac{3}{8}$- or $\frac{1}{2}$-in smooth wire
Minimum spacing	None	1 in
Maximum spacing	The lesser of: 16 bar diameters 48 tie diameters Least dimension of column	3 in

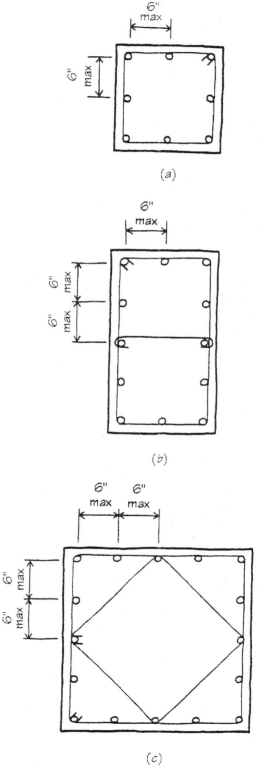

(a)

(b)

(c)

FIGURE 10.7   Typical tied column patterns.

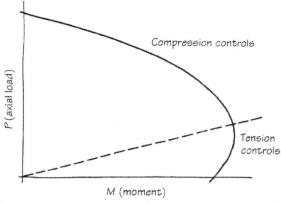

FIGURE 10.8   Typical interaction curve.

ties, respectively, for a given column. The appropriate ϕ factor is usually included in these quantities. The flat region at the top of each curve represents an ACI Code limitation on the theoretical axial capacity and ensures that a column would be designed for at least a small amount of moment. The kink in the lower end of each curve represents the fact that as the member changes from a beam-column to strictly a bending member, the ϕ factor makes the transition from 0.65 (laterally tied) or 0.70 (spirally tied) to 0.90 for bending.

Because of the many variables involved, e.g., material strengths, column dimensions, steel amount and placement, and the relative amounts of axial load and moment, it is diffi-

FIGURE 10.9   Interaction curves for design.

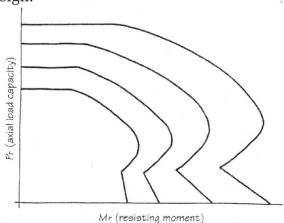

cult to provide a simplified yet realistic approach to column design. Fortunately, the level of accuracy that we need in estimating column sizes is much less than that for beams. The depth of a floor system is often crucial in helping us to compare the feasibility and attributes of one plan configuration to another, but the column dimensions seldom become a deciding factor. (This is definitely not true for very tall structures, however, where unrestrained column sizes can use up a significant amount of the floor space on lower levels, often the most attractive rental space.) It should also be noted that most inexperienced designers would, when forced to make guesses, underestimate beam depths needed for span and load conditions and overestimate column dimensions. In other words, as the structural system moves from preliminary design through final design the beams would usually get larger and the columns smaller. Suffice it to say, for a text such as this one, paying proper attention to beam behavior and design is more important than attempting a rigorous coverage of concrete columns. Consequently, only three graphs are provided in the Appendix (Graphs C.2) as aids to the preliminary sizing of laterally tied columns. The graphs assume $f_y = 60$ ksi and $f'_c = 4000$ psi and gross steel ratios ($\rho_g = A_s/bh$) of 2, 4, and 6%, respectively. The steel is assumed to be equally placed on all four sides. The moment capacity then would be for bending about either principal axis, but not both simultaneously. The following examples illustrate the use of these graphs. Once. again, the graphs are for preliminary design only; the final design mist be accomplished with more sophisticated aids or computer software.

Example 10.1   A 12 × 12-in column with four #8 bars, one in each corner, is subjected to a factored axial design load of 200 kips and a factored design moment of 55 kip-ft. If $f'_c = 4000$ psi and $f_y = 60$ ksi, will this section be adequate?

Solution: Using Table A.1 to get A_s the gross steel ratio is

$$\rho_g = \frac{A_s}{bh}$$

$$= \frac{3.16}{(12)(12)}$$

$$= 0.022$$

or just over 2%. Graph C.2 (2%) shows us that a 12 × 12-in column with a steel ratio of 2% has an axial resistance of 200 kips when the design moment is 60 kip-ft. Since the actual steel ratio is greater than 2%, the column is more than adequate.

Example 10.2   Interior columns of a 10-story building are located on a 24-ft-square grid. Assume $f'_c = 4000$ psi, $f_y = 60$ ksi, and a 4% steel ratio. Determine the required cross-sectional dimensions for a typical first floor column if the service loads at each level (floors and roof) are 60 psf of live load and 80 psf of dead load (including all self-weight). An indeterminate analysis has determined that the critical column moments are negligible; axial loads are the only design concern.

Solution: First, determine the design load for the column; then use Graph C.2 (4%) to select a preliminary column size. The factored uniform load is

$$U = 1.2(80) + 1.6(60)$$

$$= 192 \text{ psf}$$

Each level has a tributary area per column of 24 × 24 ft, or 576 ft². Therefore, each deck (floor or roof) would deliver a design load to the column of

$$576 \times 192 = 110\,600 \text{ lb} \approx 111 \text{ kips}$$

Since there are 10 decks; the total design load is

$$111 \times 10 = 1110 \text{ kips}$$

and Graph C.2 (4%) indicates that a 19 × 19-in column would be needed. (*Comment:* Most codes would permit a live load reduction in such cases on the basis that it is unlikely that the entire tributary area of all 10 levels would be loaded with the full service load at the same time. It is probable that the actual columns needed would be smaller.)

Example 10.3   A 16 × 16-in column reinforced with eight #9 bars (equally placed on all four sides) must carry a factored axial design load of 400 kips. Approximately how much moment capacity would it have?

Solution: The gross steel ratio is

$$\rho_g = \frac{A_s}{bh}$$
$$= \frac{8.00}{(16)(16)}$$
$$= 0.031$$

or about 3%, which enables us to interpolate between the two graphs, i.e., those provided for 2 and 4%. Graph C.2 (2%) indicates that when the axial load is 400 kips, the moment could be as large as 150 kip-ft, and Graph C.2 (4%) indicates an available moment capacity of about 220 kip-ft for the same axial load. Therefore, we get a design moment for our 3% + steel ratio of about 185 kip-ft.

10.6 SLENDERNESS CONCERNS

Most concrete columns tend to be fairly large in cross section compared to those of wood and steel, so buckling is of relatively less concern. The Code approach to the design of slender columns when they do occur is to apply a magnifying factor to the column moments. This factor depends upon the amount of axial load present, the loading pattern (Figure 10.2), and effective length. The effective length can be difficult to assess because it involves the amount of end re-straint afforded to the top and bottom of a column by the elements framing into it, and this is a function of the stiffness of those elements. For frames braced against sidesway the effective length is often between 0.5 and 0.9 of the column's true length, but for frames designed to permit sidesway it is almost always larger than 1.0. Figure 10.10 illustrates some effective lengths. Obviously, the greater the effective length is, the greater will be the possibility of a buckling problem, so larger

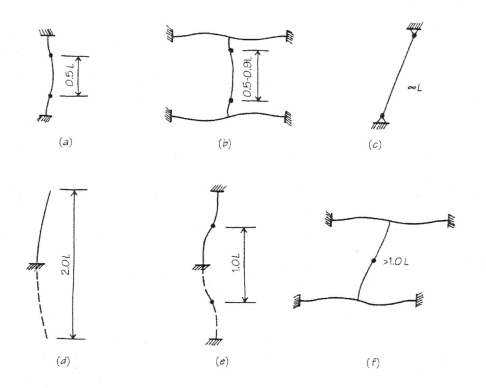

FIGURE 10.10
Effective lengths.

moment magnifiers are used (resulting in a thicker column). The proper design of a long, slender column is best accomplished with the aid of a computer because the relative size of the column influences the relative stiffness, and consequently the amount of end restraint, so an interactive process becomes inevitable.

By way of a guideline for preliminary design, columns with L/d ratios of 10 or less will probably have little or no slenderness problem. Those with an L/d between 10 and 20 will sometimes need moment magnifiers, and when L/d passes 20, buckling is quite likely to control the design.

PROBLEMS

10.1 Will a column with an 18 × 18-in section reinforced with eight #8 bars be able to sustain a compressive load of 700 kips as well as a moment of 170 kip-ft? Let $f'_c = 4000$ psi and $f_y = 60$ ksi.

10.2 For preliminary design purposes the column section for the lower floors of a 10-story building has been set at 22 × 22 in and reinforced with eight #10 bars. On the upper floors the section is to be reduced to 16 × 16 in. If $f_y = 60$ ksi, $f'_c = 4000$ psi, the area of steel used for reinforcing does not change, and the factored axial loading from the columns of the upper floors never exceeds 500 kips, will this new section be adequate to carry a new factored moment load of 150 kip-ft?

10.3 Show an acceptable lateral tie pattern for the section shown in Figure 10.11.

10.4 Referring to Graphs C.2, determine the required steel ratio for the smallest adequate square section for a column with an axial design load of 600 kips and a moment capacity requirement of 100 kip-ft.

FIGURE 10.11   Column section for Problem 10.3.

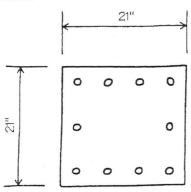

10.5 Columns of 14 × 14-in cross section set on a 30 × 30-ft grid are proposed for the third floor of a seven-story building. Service dead loads are 50 psf for floor decks and 30 psf for the roof; service live loads are 40 psf (floors) and 20 psf (roof). (Values include all selfweights.) If the gross steel ratio is 4% and the steel is evenly distributed on all sides, will an exterior column of 12-ft effective length be able to handle the moment induced by a factored wind load of 2 kips/ft? Let the moment in the column be estimated at $wL^2/10$ for wind plus a factored value of 85 kip-ft due to its continuity with the floor. Let $f_y = 60$ ksi and $f'_c = 4000$ psi.

10.6 An exterior column on the top floor of a five-story building carries a design axial load of 100 kips and a design moment load of 150 kip-ft; an exterior column on the first floor of the same building carries five times the axial load of the top floor column but has a design moment load of only 45 kip-ft. Given that the column section is to remain constant but that the steel ratio may vary, choose preliminary section dimensions and an appropriate steel ratio for each case.

11

SLAB DESIGN

11.1 THICKNESS OF ONE-WAY SLABS

The overall thickness of most one-way slabs is established by the American Concrete Institute (ACI) Code provisions for minimum thickness. In lieu of providing accurate deflection computations, the Code indicates that the minimum "thickness of construction" values of Table 11.1 may be used. For many beams the proportions dictated by good design practice and economical construction will result in thicknesses greater than these. However, for slabs this is usually not the case.

A one-way continuous slab spanning 12 ft will have the following thickness if dictated by Table 11.1:

$$h = \frac{12(12)}{28} = 5.14 \approx 5.5 \text{ in}$$

The same span if simply supported rather than continuous will have

$$h = \frac{12(12)}{20} = 7.2 \approx 7.5 \text{ in}$$

11.2 TEMPERATURE/ SHRINKAGE STEEL

Concrete elements require reinforcing to be placed for the control of cracks due to temperature and shrinkage stresses wherever reinforcing has not been placed for structural reasons. In beams and columns plenty of steel

TABLE 11.1 *Minimum thicknesses*

	Minimum Thickness, h			
Solid one-way slabs	L/20	L/24	L/28	L/10
Beams	L/16	L/18.5	L/21	L/8

exists to handle the temperature and shrinkage stresses as well as stresses from live and dead loads. However, in a one-way slab temperature steel is normally required in the direction perpendicular to the moment steel. It is usually placed between the negative and positive moment steel. The Code requires that ρ_t for temperature/shrinkage be at least as large as

0.0020 for Grade 40 steel

0.0018 for Grade 60 steel

where ρ_t is based upon the full slab thickness

$$\rho_t = \frac{A_s}{bh} \qquad (11\text{-}1)$$

The Code also requires that bars placed for temperature or shrinkage be no farther apart than the smaller of five times the slab thickness, or 18 in.

11.3 STEEL REQUIREMENTS FOR MOMENT

Knowing that the primary reinforcing steel for slabs is often #4 or #5 bars and the cover requirement for slabs is ¾ in clear, it is easy to get an effective depth value. If we assume the use of #4 bars, e.g., in a slab which is 5 in thick, the d value will be 5 in less ¾-in cover less ¼-in bar radius, or 4 in net.

For continuous slabs the ACI coefficients discussed in Chapter 4 are particularly useful. For example, if we can ascertain that the slab in question fits one of the cases in Figure 4.18b, and the design moments can thus be determined, then finding out the amount of steel needed at each crucial moment location is not difficult.

Tables B.2 of the Appendix provide M_r values for a one-foot width of slab for various ρ and d combinations. Three concrete strengths for each of the two commonly used

steel strengths are included. The tables have ρ values below ρ_{min} for beams because for slabs the minimum amount of flexural steel required is governed by temperature and shrinkage rather than bending, except that moment bars shall be placed no farther apart than the lesser of three times the slab thickness, or 18 in.

Once the ρ needed has been established, it is easy to determine the spacing required to provide the corresponding amount of steel.

Example 11.1   Assume that by using the minimum thickness requirements a certain slab must be at least 6 in thick. If $f_c' = 3000$ psi and $f_y = 40$ ksi, determine the required spacing of #4 bars in order to carry a factored load moment of 4.5 kip-ft.

Solution: The effective depth will be 6 in less ¾ in less ¼ in, or 5 in. From Table B.2(40/3), for Grade 40 steel and 3000 psi concrete we find that a p value of 0.0055 is needed for an M_r of 4.5 kip-ft.

Since for a 12-in width

$$A_s = \rho(12)(d)$$

and knowing that the number of bars in a 12-in width is 12 ÷ s, then

$$A_s = \frac{12}{s}(a_s)$$

where a_s = area of one bar.

Setting these two expressions equal to one another gives us

$$s = \frac{a_s}{\rho d} \qquad (11\text{-}2)$$

Since the area of one #4 bar is 0.20 in², we get

$$s = \frac{0.20}{0.0055(5)}$$

$$= 7.3 \text{ (rounding down)} \approx 7 \text{ in}$$

This is within the Code controls of $3h$, or 18 in, for maximum spacing of moment bars.

Finally we need to check this result against the requirement for temperature/shrinkage. This is based upon the full thickness, and for Grade 40 steel we must provide a ρ_t of at least 0.0020.

$$s = \frac{a_s}{\rho_t h} \qquad (11\text{-}3)$$

$$= \frac{0.20}{0.0020(6)}$$

$$= 16.6 \approx 16.5 \text{ in}$$

Since 7 in is less than 16.5 in, the temperature/shrinkage requirement will not govern.

Temperature/shrinkage steel *will* be required at right angles to the moment steel, of course, and if #4 bars are used, they should be spaced no farther apart than 16.5 in as computed previously. This will also meet the Code maximum of $5h$, or 18 in.

Example 11.2   Determine the required depth and specify the required steel for the three-bay, one-way slab of Figure 11.1. Use f_y = 40 ksi steel and 3000 psi concrete. Use #4 bars for all steel parallel to the span and #3 bars for temperature steel in the other direction. The live load is 100 psf and the only dead load is the slab itself.

Solution: The required thickness from Table 11.1 will be

$$h = \frac{12.5(12)}{28} = 5.36 \approx 5.5 \text{ in}$$

FIGURE 11.1   Three-bay, one-way slab.

Knowing that the required cover is ¾ in and using #4 bars, we find that the effective depth will be 4.5 in. The self-weight can be determined as

$$\frac{5.5}{12}(150) = 69 \text{ psf}$$

Therefore, the factored uniform load is

$$w_u = 1.2(69) + 1.6(100)$$

$$= 243 \text{ plf} = 0.243 \text{ klf}$$

Using the ACI coefficients of Figure 4.18b, we see that the tabular format in Table 11.2 is convenient to find the required moment steel. Note that only the larger of the two moments c and d will be used since the two locations are reinforced by the same steel bars running across the beam. Also, notice that in compliance with the proper use of the coefficients the average of the two span lengths was used in determining that moment.

The Code stipulates a maximum spacing of $3h$, or 18 in for moment steel. The value of $3h$ in this case is 16.5 in. Thus, the spacing at location a should be 16.5 in unless temperature and shrinkage requirements govern.

Checking the temperature/shrinkage requirement in the direction of the span, and using #4 bars and Grade 40 steel, we get

$$s = \frac{a_s}{\rho_t h} \qquad (11\text{-}3)$$

$$= \frac{0.20}{0.0020(5.5)}$$

$$= 18.2 \approx 18 \text{ in}$$

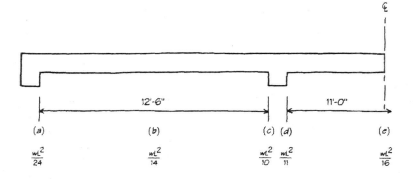

TABLE 11.2

	(a)	(b)	(c)	(e)
	$\dfrac{wL^2}{24}$	$\dfrac{wL^2}{14}$	$\dfrac{wL^2}{10}$	$\dfrac{wL^2}{16}$
	$\dfrac{0.243(12.5)^2}{24}$	$\dfrac{0.243(12.5)^2}{24}$	$\dfrac{0.243(12.5)^2}{24}$	$\dfrac{0.243(12.5)^2}{24}$
M_u (kip-ft)	1.58	2.71	3.55	1.83
ρ [Table B.2(40/3)	0.0025	0.0040	0.0055	0.0030
$s = \dfrac{a_s}{\rho d}$	$\dfrac{0.20}{0.0025(4.5)}$	$\dfrac{0.20}{0.0040(4.5)}$	$\dfrac{0.20}{0.0055(4.5)}$	$\dfrac{0.20}{0.0030(4.5)}$
s (in)	17.8	11.1	8.1	14.8
	≈ 17.5	≈ 11	≈ 8.0	≈ 14.5

for temperature steel the maximums are $5h$, or 18 in, but in this case moment controls throughout. Using #3 temperature steel in the other direction as specified in the problem statement, we obtain

$$s = \frac{0.11}{0.0020(5.5)}$$

$$= 10 \text{ in}$$

Figure 11.2 shows how the reinforcing pattern might be established. The Code requires that at least one-third the positive moment steel in simple beams and one-fourth the positive moment steel in continuous beams be carried into the support. Since the bars cannot be spaced farther apart than 18 in, it is logical to change the 11-in spacing to 9 in for the positive moment steel in the outer spans so that alternate bars could be continued into the support. All of the positive moment steel in the middle span should be continued to the beam. The reader is referred to the Code directly for the precise details of steel placement.

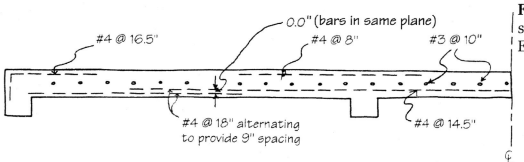

FIGURE 11.2   Slab steel requirements for Example 11.2.

FIGURE 11.3 Section through the middle of a two-way slab system; columns shown are in the background.

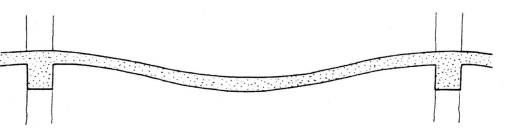

11.4 BEHAVIOR OF TWO-WAY SLABS

The proper analysis of the forces in two-way slabs is not treated in this basic text and only a brief discussion involving behavior and procedures is included here.

One-way slabs are used when the spacing of beams is relatively small, say, 6 to 16 ft, and the slab panels themselves are almost always rectangular. The slab action carries the loads to the beams, which in turn deposit their reactions as concentrated loads on girders (Figure 3.6).

Two-way action develops in slabs that are more square where the beams frame directly into columns at the four corners of the slab (Figure 3.4). For smaller spans the beams can be omitted and the system is then more properly called a flat plate (Figure 3.2). Flat plates and two-way flat slabs behave in much the same manner, except that plate systems have a problem with shear around the base of the columns (punching shear). For this reason flat plates often have a thickened portion in the immediate vicinity of the columns in the form of a *drop panel* or conical *shearhead* (Figure 3.3).

Deflection often controls the thickness of two-way slabs, just as with one-way slabs, but the rules for minimum thickness are not nearly so simple. A two-way slab dishes downward in the middle such that a midspan section cut through in either direction would appear as in Figure 11.3. (The beam deflections would be relatively small and for clarity are not shown.)

FIGURE 11.4 Two-way flat plate.

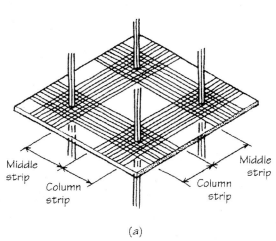

Middle strip

Column strip

Column strip

Middle strip

(a)
column strips and middle strips

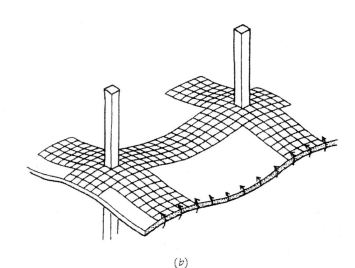

(b)
distribution of positive moment

The reinforcement, of course, runs in both primary directions and is located near the bottom of the slab in areas of positive moment (the "dish") and near the top of the slab in areas of negative moment, where the slab passes over beams. This is true whether or not beams actually exist because in their absence the portions of the plate that connect the columns tend to behave like wide shallow beams (see Figure 11.4). These form what are called *column strips*.

For analysis purposes the slab is divided into an orthogonal grid of *column strips* and *middle strips*. Column strips tend to be stiffer than middle strips and consequently have larger moments. When the column strip contains a beam (flat slab system), most of the moment is carried by the beam. As with most continuous structures, the larger moments tend to be the negative ones near the supports.

One popular simplified method of analysis assumes that the moment per foot of width is constant over the width of each strip and assigns positive and negative moment coefficients to be used to determine the required spacing of steel for each location. The minimum amount of steel in each direction is that required by temperature and shrinkage considerations with the Code placing a limit on the maximum spacing of bars as the smaller of twice the slab thickness, or 18 in.

11.5 SLABS-ON-GRADE

A slab placed on compacted soil or on a specially prepared subbase of sand or gravel is called a *slab-on-grade*. It theoretically needs reinforcement only for the control of cracking due to shrinkage and temperature stresses. However, most designers recognize that due to uneven loading and the practical difficulties of obtaining a uniformly stiff base, the steel is useful for the control of cracks caused by bending and shear as well.

Such slabs vary in thickness from a minimum of 4 in for residential applications to as much as 8 in or more for heavy warehouse loadings. For thinner slabs one layer of welded wire fabric (usually a 6-in mesh) is placed about 2 in below the top. For thick and heavily loaded slabs two layers of mesh or grids of reinforcing bars are used, located 2 in from the top and bottom surfaces.

In the case of soils which are very nonuniform in their makeup or subject to large volume changes posttensioning is sometimes used to help maintain the integrity of the slab.

PROBLEMS

11.1 A one-way slab simply spans 11 ft and is 7 in thick. The factored loads (including selfweight) total 250 psf. Assuming Code standards and limits and letting $f_y = 60$ ksi and $f'_c = 3000$ psi, determine the proper spacing of reinforcement for moment (use #4 bars) and temperature/shrinkage (use #3 bars).

11.2 A simply supported, one-way slab spanning 15 ft is reinforced for a moment with #4 bars spaced 10 in apart. The slab is 9 in thick. What is the maximum uniform service live load that this slab can support if all Code requirements are met, the steel is 60 grade, and the concrete strength is 4000 psi? Assume there is no *applied* dead load.

11.3 Propose reinforcement for a simply supported, one-way slab that spans 12 ft carrying a live load of 95 psf and an applied dead load of 70 psf. Assume 4000-psi concrete; use 60-ksi #5 bar for moment and 40-ksi #4 bar for temperature steel.

11.4 A cantilevered floor slab extending 8 ft from a fixed support must carry applied loads of 50 psf (live) and 20 psf (dead). Determine reasonable thickness and effective depth and specify adequate reinforcement for the slab using 60-ksi steel and assuming 3000-psi concrete. Use #4 bar in the direction of the cantilever and #3 bar for temperature/shrinkage steel.

11.5 The four-span continuous slab shown in Figure 11.5 carries a live load of 120 psf and a dead load consisting only of its own weight. Using 60-ksi #5 bar for moment and 40-ksi #3 bar for temperature/shrinkage steel, propose reasonable reinforcement spacings for the entire slab. Let $f_c = 4000$ psi. Provide an answer sketch similar to Figure 11.2.

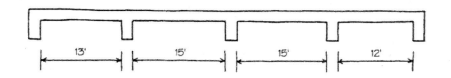

FIGURE 11.5 Slab for Problem 11.5.

12

FOUNDATION SYSTEMS

12.1 SOIL PROPERTIES

Soil is often classified by its constituents. Table 12.1 provides some commonly accepted names and their approximate range in terms of particle size.

The distinctions are not clear cut and there is overlap in the size ranges between sand and silt and between silt and clay. The names are based upon characteristics and behavior rather than upon size.

A laboratory analysis of a soil will often include a sieve analysis in which air-dried pulverized soil samples are shaken through a series of sieves and the percent retained on each sieve is recorded. The approximate opening size of the finest sieve (a #200) is 0.003 in and particles which pass through this are called *fines*; the remaining particles are called *coarse*. The *Unified Soil Classification System* classifies a soil as coarsegrained if more than 50% of the particles are retained on the #200 sieve and as finegrained if 50% or more pass through the sieve.

Soils are also classified as *cohesionless* and *cohesive*. The coarse-grained soils tend to be cohesionless whereas the fine-grained soils tend to be cohesive. A cohesionless soil falls apart when dry (sand) whereas a cohesive soil

TABLE 12.1 *Sizes of soil particles*

Particle Type	Approximate Diameter (in)	
Cobble	> 3.0	
Gravel	0.25 – 3.0	Coarse
Sand	0.003 – 0.25	
		# 200 Sieve
Silt	0.0001 – 0.003	Fine
Clay	< 0.0001	

tends to stick together when dry (clay). Cohesionless soils carry loads by the development of frictional forces between the particles and cohesive soils carry loads by the development of shearing and tensile stresses. Cohesionless soils, particularly the coarse sands and gravels, are less affected by the presence of moisture than are cohesive soils. (A fine sand, however, can become "quick" with the addition of water, which serves to reduce the frictional forces.) Cohesive soils are usually very sensitive to water and their primary characteristics are heavily dependent upon moisture content.

A typical soil contains a mixture of sand, silt, and clay, so its behavior and suitability as foundation material will depend upon the nature of the individual constituents and the fraction of each. In general, coarse-grained soils have greater bearing capacities than fine-grained soils. Also, in general, bearing capacities increase with the depth below the surface, and this effect is more pronounced in cohesionless soils because the frictional forces increase with depth. A proper determination of the safe bearing capacity can only be done by:

1. Drilling to determine the types of soil present

2. Performing field tests such as the *Standard Penetration Test*, which provides a measure of the density or stiffness of the subsurface layers

3. Performing laboratory tests to establish the soil characteristics

12.2 SHALLOW FOUNDATION SYSTEMS

When favorable soil strata are located at the surface and when the building is not too heavy, a shallow foundation system consisting

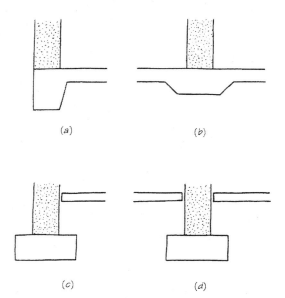

FIGURE 12.1   Exterior and interior wall footings.

of spread footings is the logical choice. The concept is very simple; the footing merely provides additional bearing area below the wall or column so that the actual pressure is less than the allowable soil bearing capacity. In Figure 12.1 footings *a* and *b* would be suitable for light slab-on-grade construction; however, in most construction it is desirable to isolate the floor slab from the footings as shown in Figure 12.1*c* and *d*. The latter approach permits movement of one relative to the other and helps to prevent the slab from cracking.

In the case of exterior footings it is imperative that the base of the footing be placed below the frost depth for that locality. In colder climates this depth can be considerable and provides the rationale for creating basement space.

When two column footings are close together, they may be combined into one; the area between the columns would then have to be designed as a beam with an upward acting load from the soil. The use of a *combined footing* is particularly appropriate when the building is close to the property line where the outer column would otherwise have to load its footing eccentrically.

As illustrated in Figure 12.2a, the shape of the combined footing should reflect the different column loads so their resultant passes through the centroid of the footing. When the distance between the two footings becomes larger, an actual beam (called a *strap*) is used to connect the two pads.

The pressure distribution below a spread footing is called a pressure bulb and its shape is dependent upon the size of the footing and the soil characteristics. Pressure contours as shown in Figure 12.3 illustrate that the pressures (as one would expect) get smaller as the distance from the base of the footing increases. The pressure contours also show the fluidlike nature of the soil in that these pressures act laterally as well as downward; near the edge of the footing there is even an upward pressure!

Figure 12.3 also shows that, given the *same magnitude of pressure* at the base of the footing, wall footings have deeper pressure bulbs than column footings because of the overlapping pressures that occur, apparent if one thinks of a wall as a series of adjacent columns. Larger footings required for larger column loads also stress soil at greater depths.

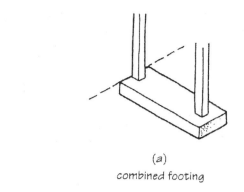

(a)
combined footing

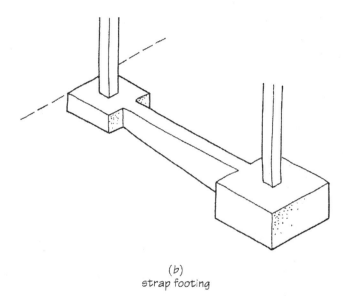

(b)
strap footing

FIGURE 12.2 Footings at the property line.

FIGURE 12.3 Typical pressure bulbs where 1.0 is the pressure at the base of the footing.

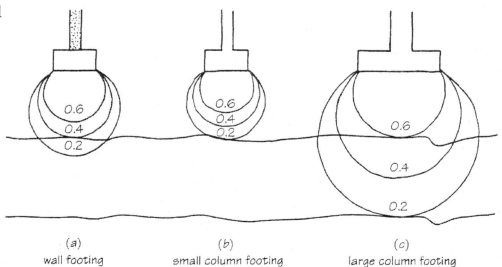

(a)
wall footing

(b)
small column footing

(c)
large column footing

This is important to remember in situations such as that shown where a layer of inadequate soil exists several feet below the surface. The result could be significant differential settlement from footing to footing.

Even though layers of poor soil are encountered beneath the surface, soils usually increase in bearing capacity with depth, and in many situations adequate soil for spread footings can be found by excavating a few more feet. This assumes that the presence of groundwater does not interfere with construction. (*Dewatering* or lowering the water table at a building site can be expensive.) Another reason for going deeper may be to achieve an increased resistance to uplift. It is important to remember that the foundation system must also resist lateral forces which tend to cause sliding and overturning.

Soils of low bearing capacity require large footings, and if the sum of the areas of individual footings gets up as high as 40 or 50% of the plan area, then it may be more economical to consider a *mat* foundation. In this case all the columns share one large footing that covers the entire plan area. Such footings are also useful when there are large variations of soil capacities on a given site. Figure 12.4 shows variations of mat footings with increasing column loads from *a* to *c*.

When a mat is placed deep in the soil such that the weight of the excavated earth equals most or all of the building weight, it is called a *raft* or *compensated* footing. It is also sometimes referred to as a *floating* foundation, but this term should be reserved for cases where a watertight compensated system is used below the water table.

12.3 DEEP FOUNDATION SYSTEMS

When the soil near the surface does not have enough bearing capacity to use spread footings, a deep foundation system using piles

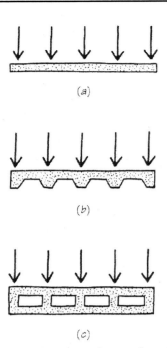

FIGURE 12.4   Types of mat footings.

must be considered. Inevitably more expensive, piles must be used for most tall buildings because of weight and stability considerations. Piles can be made of either timber, concrete (precast or cast-in-place), or steel, or combinations of these materials. Larger isolated cast-in-place concrete piles are sometimes called *piers* or even *caissons*, but the latter term should be reserved for underwater applications.

Piles are often classified as *friction* piles and *point* or *end-bearing* piles, and the two types are illustrated in Figure 12.5. It should be noted that the shape of the pressure bulb around a friction pile is heavily dependent upon the soil properties. In actuality, many piles act through a combination of these two effects. Friction piles are usually tapered to facilitate driving and to increase the downward load capacity. End-bearing piles are usually driven through softer material to a very stiff or impenetrable base. The problem then is to make sure the pile can resist buckling if the soft material cannot provide adequate lateral support.

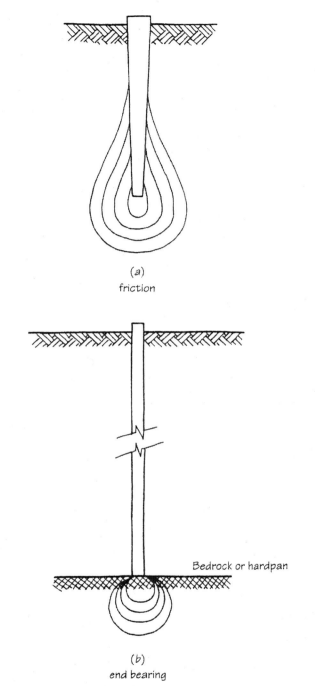

(a)
friction

(b)
end bearing

Bedrock or hardpan

FIGURE 12.5 Pile types.

All piles with the exception of those cast-in-place are subject to lateral displacement by nonuniform soil conditions and damage by the forces of driving or from hitting boulders. For these reasons large factors of safety are

often used in pile design, and piles are used in groups or clusters of three or more to achieve the required load capacity. A very stiff *pile cap* of cast-in-place concrete is needed to attempt to distribute the column load uniformly to all the piles in a group. It should be noted that a large group of piles has a much deeper pressure bulb than that of a small group.

When piles are driven at an angle to the vertical for the purpose of resisting lateral forces, they are called *batter* piles. These were once used only for bridge and overpass construction but are becoming increasingly common for buildings as well.

Timber piles have the least bearing capacity but also the lowest initial cost. They are frequently 25 to 40 ft in length but can be much longer depending upon the availability of tall trees. They must always be treated with a preservative because, although they will not rot below the water table, they will suffer deterioration due to continuous wetting and drying above it. They vary in safe bearing capacity from about 30 to 80 kips.

Precast concrete piles are often prestressed to help cope with driving and handling stresses. These piles are usually 30 to 70 ft long but can be much longer if the equipment is available for handling them. They are most useful in uniform soil conditions where the depth at which adequate resistance will be found can be predicted because they are difficult to cut off and very difficult to splice (if additional depth is needed). Their load capacity varies widely but is typically in the 70- to 150-kip range.

A popular concrete pile system involves the driving of a (smooth or corrugated) *steel shell*, which is then *filled with concrete*. This type of pile has the significant advantage that the shell can be inspected for straightness before filling with concrete. The length range is extensive, from 30 to about 120 ft, with the longer shells having steps or shoulders to provide more places for the driving shaft (or *mandrel*) to impact the shell as in Figure 12.6*d*. Depending upon their length and size, these "shelled" piles can support 80 to 160 kips. A more economical pile results if the shell can be removed before the concrete is placed. The first batch of concrete can be compacted by ramming and is actually forced outward at the bottom of the pile as in Figure 12.6*b*. The removal of the shell assumes that the surrounding soil is stable enough to maintain its shape while the concrete is placed and hardens.

Cast-in-place concrete piers are sometimes used for heavier buildings. They may or may not make use of a belied-out bottom as seen in Figure 12.6*a*. The "bell" provides additional bearing area and greatly increased uplift resistance. Such piers are the only feasible system where good bearing capacity lies deep below layers of weak or unstable soil such that slender (steel) piles would buckle. (Many of the tall buildings in Chicago utilize large piers of this type.) In some cases the walls of the pier must be temporarily lined with steel to prevent collapse or the hole filled with a bentonite slurry which is later displaced as concrete is placed by tube from the bottom up. Modern rigs are capable of drilling holes up to 12 or 14 ft in diameter and more than 100 ft deep. The capacity of such piers is measured in hundreds of kips.

Steel piles of pipe or H-shapes are often used for point bearing applications with heavier loads. They are the easiest to splice if additional length is needed; usually 60 to 120 ft long, they have been driven to depths of over 250 ft. Their small cross section makes them easier to drive than other piles but also more subject to unknown displacement by boulders or other obstructions. Their load range varies from 80 to 300 kips or more.

The preceding survey of pile systems is by no means exhaustive and there are many combinations and variations in use today.

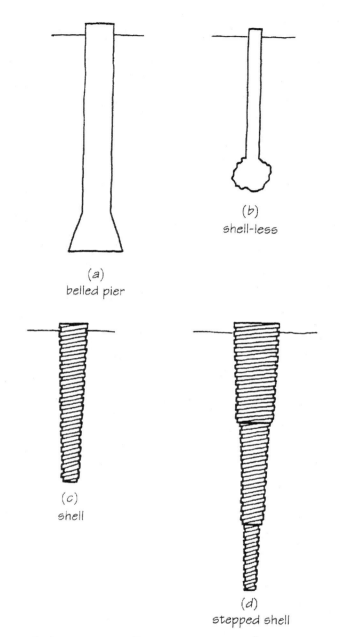

FIGURE 12.6 Common types of concrete piles.

Composite piles using timber or steel below the water table and concrete above makes greater depths more economical. Noise pollution ordinances can make drilled piers or vibration-assisted driving systems more popular. There is a considerable amount of developmental activity in this area of construction and many of the latest techniques are only available on a local or regional basis. The designer of a building utilizing a deep foundation system is urged to obtain the best advice available in terms of local engineering and construction expertise.

12.4 DESTRUCTIVE SETTLEMENT

Buildings are expected to settle slightly and little can be done to prevent some movement from occurring. Cohesionless soils tend to settle more or less instantaneously whereas some cohesive soils are very time-dependent. Uniform settlement is seldom a problem. Differential settlement, however, may cause finished surfaces to crack, floors to slope, and doors and windows to jam, and add stress (sometimes dangerously so) to load-carrying members. Figure 12.7 serves to explain why minor settlement differences would have a much greater effect upon the structural integrity of indeterminate or continuous structures

than upon determinate ones. For the rigid frame of Figure 12.7b the negative moment at the left end of the beam would be *additive* with that from the gravity load. Furthermore, the stiffer the frame is, the more sensitive it would be to small displacements.

The most common cause of differential settlement is nonuniform soil conditions, some of which are illustrated in Figure 12.8. These provide more than adequate reason to insist upon a thorough investigation of any potential building site. The cost of not taking enough test borings or not sampling deep enough could be very high indeed.

FIGURE 12.8   Soil conditions which can cause differential settlement.

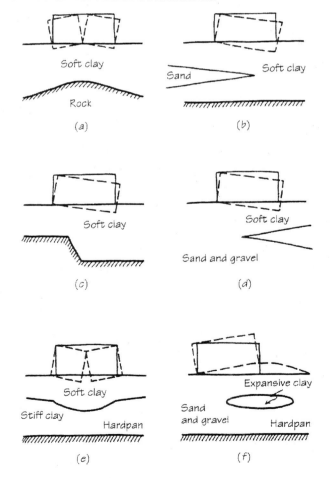

FIGURE 12.7

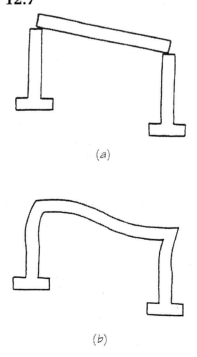

Some further causes of destructive settlement are:

Shallow systems

1. Eccentric loading conditions
2. Adjacent new construction that results in:
   a. Pressure on the top of an existing footing
   b. Overlapping pressure bulbs
   c. Settlement that causes a pressure bulb to change shape
   d. Excess vibration
3. Forgetting that a slab on grade will load the footings below
4. Rapidly growing vegetation that changes the soil's moisture content
5. Frost heave

Deep systems

1. Damaged or broken piles not discovered during driving
2. Deterioration of wood piles
3. Voids in cast-in-place concrete piles (air, intervening soft clay)
4. Adjacent new construction that results in:
   a. Overlapping pressure bulbs
   b. Negative friction on existing piles c. Excess vibration
5. Forgetting the large pressure bulb associated with a large pile group

Settlement is also very much related to the proportion of dead and live loads that act on various footings. The dead load is there all the time and is more likely than live load to cause an increase in settlement with time. Codes recognize that for members with large tributary areas, it is unlikely that every square foot will be loaded with the full live load at the same time, and reductions in live load are usually permitted in such cases. The foundation for a building of more than one or two stories then will be designed for all the dead load and a portion of the live load. Designers recognize that although this provides adequate protection against failure, it does not treat the issue of differential settlement very well. It is also prudent to proportion the areas of various footings so that equal soil pressure is present due to the loads likely to be sustained most of the time. This is especially true in situations where the ratio of live load to dead load changes markedly from footing to footing. One can understand the complexities that result when the soil bearing capacity also varies from footing to footing.

13

FOOTING DESIGN

13.1 INTRODUCTION

Footing design is basically a three-stage process. First, the overall required bearing area is determined such that the allowable soil bearing capacity is not exceeded. Then the necessary footing thickness is computed such that the concrete will not fail in shear. Finally, reinforcing is placed for bending and for temperature and shrinkage.

Allowable bearing pressures are based upon soil tests and are used to prevent the footing from punching down into the soil and to avoid excessive settlement. A factor of safety in the range of 2.5 to 3.0 is used based upon the actual soil strength. A large factor of safety is prudent since soils can vary considerably and the performance of the overall building is heavily dependent on that of the foundation system. Local building codes will sometimes be able to provide tables of predicted safe bearing capacities for the various soils in a particular area.

In determining the overall plan dimensions of a footing (the width in the case of a wall footing), we need to take into account *all* of the loads which contribute to the pressure upon the soil underneath the footing. These may include:

1. The actual dead and live loads applied to the footing via the wall or column

2. The weight of the footing itself

3. The weight of the soil above the footing

4. The weight of the slab-on-grade (if any) and the loads acting on it

The loads in this case should be the unfactored service loads since we are working with *allowable* soil capacities. To determine the required footing thickness and the appropriate amount of steel, we use strength design procedures with factored loads.

Footings tend to be quite thick since the American Concrete Institute (ACI) Code requires a minimum of 6 in of concrete above the reinforcing and 3 in of cover below. The reinforcing itself requires some space, so the end result is that footings are seldom less than about one foot thick.

13.2 COLUMN FOOTINGS

The thickness of a typical square column pad is almost always governed by *punching shear*, a type of failure in which the column pushes through the footing. From another standpoint one could also say that the soil pushing upward fractures the pad so that a square ring or "donut" of footing slides upward relative to the base of the column. This type of failure results in the 45° pyramidal shape shown in Figure 13.1a. To avoid the unnecessary complication of having to calculate the shear stresses on these sloped surfaces, it is common practice to work with the fictitious vertical failure surfaces of Figure 13.1b. These vertical surfaces are located $d/2$ beyond the face of the column to approximate better the surface area of the actual failure. The perimeter of the assumed failure surface is four times one of its sides, or

$$4 \times (\text{column width} + d)$$

and is called b_o. The subscript here implies distance "around," or perimeter.

The Code provides the strength of concrete in punching shear as $4\sqrt{f'_c}$, where the units are in psi. This is twice the shear strength given for beams in Chapter 8 and is appropriate because the local compressive stresses at the base of a column serve to increase the shear strength. Using the capacity reduction factor, the punching shear strength is then given as

$$V_c = \phi 4\sqrt{f'_c} b_o d \qquad (13\text{-}1)$$

where ϕ is 0.75 for shear.

The effective depth d is used rather than the full thickness and any contribution to shear resistance provided by the moment steel is ignored.

The thickness is determined such that the shear force resulting from the factored loads does not exceed that given by Equation (13-1). It is usually not feasible to provide special reinforcing to reduce the thickness of footings. (However, this is sometimes done to control punching shear around the columns of flat plate floor decks. As mentioned in Chapter 11, the thickness of the floor is frequently increased adjacent to the columns to prevent punching shear.)

Once the footing thickness is known, steel is selected to carry the moment. The "beam" in this case is each of four parts of the footing that project beyond the faces of the column like wide inverted cantilevers (Figure 13.2).

FIGURE 13.1 Punching shear failure, actual and assumed.

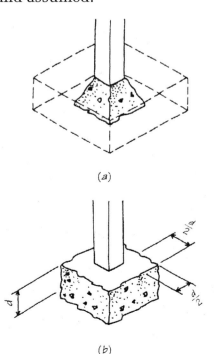

(a)

(b)

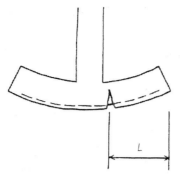

FIGURE 13.2

The critical section exists at the column face, so the length of the cantilever will be $L = \frac{1}{2}$(footing length – column width).

    Because the "beam" has a relatively large d, usually the steel requirement will be low. Sometimes ρ_{min} will control. Since a footing acts like a two-way slab, ρ_{min} will be 0.0020 for F_y60 steel and 0.0018 for F_y40 steel.

    In the case of a column pad no steel will be needed for temperature and shrinkage since the moment steel runs both ways. In this regard it should be noted that such footings will always be slightly stronger in bending in one direction than the other because the two d distances cannot be the same. The lesser of the two should be used in determining the proper amount of steel. The difference between d and h is often 5 or 6 inches.

    If space limitations dictate the use of a less economical rectangular footing, the design procedures will be the same except that a second type of shear failure may govern the thickness, i.e., *beam shear* or so-called *one-way shear* across the narrow dimension. This type of failure and its assumed counterpart are shown in Figure 13.3. just as in a beam, the failure plane is located d distance from the face of the column. The appropriate shear strength is

$$V_c = \phi 2\sqrt{f_c'}\,bd \qquad (13\text{-}2)$$

where b is the footing width.

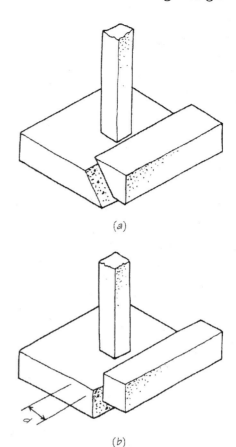

FIGURE 13.3   Beam shear failure, actual and assumed.

Example 13.1   Design the square column pad of Figure 13.4. The dead load, including the column itself, is 100 kips and the live load is 85 kips. The weight of the soil is 100 pcf and its allowable bearing capacity is 3500 psf. Use $f_c' = 3000$ psi and $f_y = 60$ ksi.

Solution: Assume a trial thickness of 18 in; depending upon the size of the reinforcing bars, this means a minimum d distance of about 13 in. An effective allowable bearing pressure q_e can be obtained by subtracting from the given 3500 psf all the loads present which are not part of the column load, in this case 1½ ft of concrete and 2½ ft of soil.

$$q_e = 3500 - 1.5(150) - 2.5(100)$$

$$= 3025 \text{ psf} \quad \text{or} \quad 3.025 \text{ ksf}$$

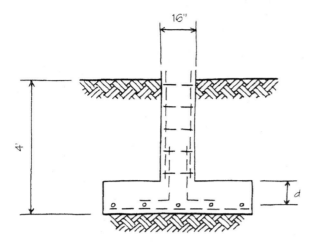

FIGURE 13.4

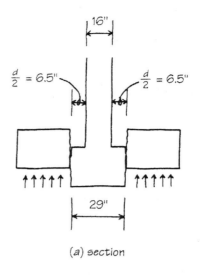

(a) section

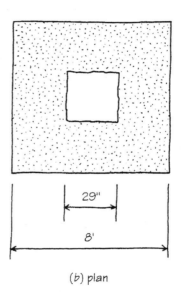

(b) plan

FIGURE 13.5

The required area of footing is

$$A_r = \frac{P}{q_e}$$

$$= \frac{100 + 85}{3.025}$$

$$= 61.1 \text{ ft}^2$$

Plan dimensions are often done in 3-in increments, in this case resulting in an even 8-ft dimension so the provided area will be 64 ft².

Now the punching shear can be checked to see if the trial thickness of 18 in is enough. The factored column load is

$$P_u = 1.2(100) + 1.6(85)$$

$$= 256 \text{ kips}$$

(Note that the column load is the only one that will cause shear and moment in the pad; the weight of the footing and the soil above will not.)

The upward pressure on the base of the footing due to the factored load is

$$q_e = \frac{P_u}{A}$$

$$= \frac{256}{64}$$

$$= 4.0 \text{ ksf} \quad \text{or} \quad 4000 \text{ psf}$$

The punching shear force due to this load can be calculated with the aid of Figure 13.5. Only the pressure on the shaded area (the "donut") causes punching shear:

$$V_u = 4000\left[8^2 - \left(\frac{29}{12}\right)^2 \right]$$

$$= 233\,000 \text{ lb}$$

The punching shear strength is, according to the Code,

$$V_c = \phi 4\sqrt{f_c'}b_o d \qquad (13\text{-}1)$$

The perimeter of the hole is 4×29, or 116 in:

$$V_c = 0.75(4)\sqrt{3000}(116)(13)$$

$$= 248\ 000\ \text{lb}$$

Since $V_c > V_u$, the thickness is adequate and is slightly overdesigned.

Now determine the amount of steel needed for the moment. The length of the cantilever is $\left(8 - \frac{16}{12}\right)/2$, or 3.33 ft, and its width is 8 ft

FIGURE 13.6

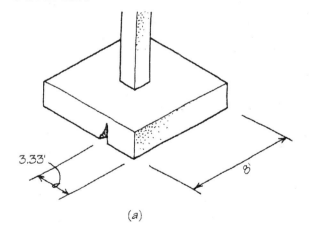

(a)

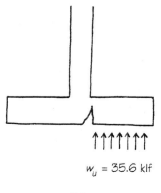

$w_u = 35.6$ klf

(b)

(Figure 13.6). The design load per foot will equal the pressure times the width:

$$w_u = q_u(8)$$

$$= 4.0(8)$$

$$= 32.0\ \text{klf}$$

So, the moment at the back end of the cantilever is

$$M_u = \frac{w_u L^2}{2}$$

$$= \frac{32.0(3.33)^2}{2}$$

$$= 177\ \text{kip-ft}$$

We must provide at least this much resisting moment:

$$\frac{M_r}{\phi b d^2} = R \qquad (7\text{-}2a)$$

$$\frac{177(12)}{0.9(96)(13)^2} = 0.145$$

Using Table B.1(4), we find that $\rho = 0.0025$.

$$A_s = \rho b d \qquad (6\text{-}7)$$

$$= 0.0025(96)(13)$$

$$= 3.12\ \text{in}^2$$

Table A.1 indicates that eight #6 bars will provide 3.52 in². The footing will need this much steel in each direction, of course. Sometimes a smaller number of large bars cannot be used because of the maximum spacing requirement of 18 inches.

13.3 WALL FOOTINGS

The approach to design for a wall footing is very similar to that for a column pad except,

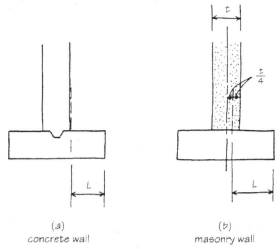

(a)
concrete wall

(b)
masonry wall

FIGURE 13.7   Critical sections for moment.

of course, that the four-sided punching shear failure cannot take place. A one-foot length of wall is usually analyzed and the previously mentioned one-way or beam shear governs the thickness if the Code minimum requirements do not. Transverse steel is needed for moment, except in lightly loaded residential footings, and temperature/ shrinkage steel in the direction parallel to the wall is always necessary. The critical section or moment depends upon the relative stiffness of the wall and is illustrated in Figure 13.7.

FIGURE 13.8

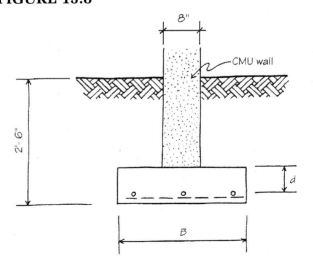

Example 13.2   Design the wall footing of Figure 13.8. The dead load, including the wall weight, is 4 kips/ft and the live load is 3 kips/ft. The soil weight is 110 pcf and its bearing capacity is 2750 psf. Use $f_c' = 3000$ psi and $f_y = 40$ ksi.

Solution: Assume a trial thickness of 12 in which, with the required 3-in cover, will give us a d value of about 8.5 in. The effective allowable bearing pressure will be

$$q_e = 2750 - 1.0(150) - 1.5(110)$$

$$= 2435 \text{ psf}$$

The required width of footing B will be

$$B_r = \frac{P}{q_e}$$

$$= \frac{4000 + 3000}{2435}$$

$$= 2.87 \text{ ft}$$

Using a 3-ft wide footing, the factored loads will provide an upward pressure on the base of the footing of

$$q_u = \frac{1.2(4) + 1.6(3)}{3.0}$$

$$= 3.20 \text{ ksf} \quad \text{or } 3200 \text{ psf}$$

FIGURE 13.9

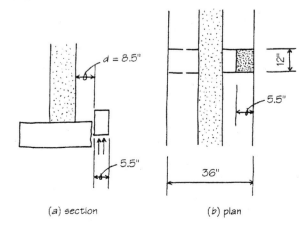

(a) section

(b) plan

Now we can compute the shear force to check the adequacy of our assumed thickness. Moving out from the face of the wall a distance d and referring to Figure 13.9, the shear force will be

$$V_u = 3200\left(\frac{5.5 \times 12}{144}\right)$$

$$= 1467 \text{ lb}$$

The beam shear strength is, according to the Code,

$$V_c = \phi 2\sqrt{f_c'}bd \qquad (13\text{-}2)$$

$$= 0.75(2)(\sqrt{3000})(12)(8.5)$$

$$= 8380 \text{ lb}$$

Since $V_c \gg V_u$, the thickness is too great. We reduce it and try again. Bearing in mind that 6 in of concrete is required above the steel and 3 in below, the minimum thickness will be about 10 in. (We hold the width at 3 ft as the effective bearing capacity will increase only slightly with the reduction in footing thickness.) The new d will be about 6.5 in and the new V_u becomes

$$V_u = 3200\left(\frac{7.5 \times 12}{144}\right)$$

$$= 2000 \text{ lb}$$

The new V_c then will be

$$V_c = 0.75(2)(\sqrt{3000})(12)(6.5)$$

$$= 6410 \text{ lb}$$

So, we are still much more than adequate.

The moment will need to be checked. The cantilever length, shown in Figure 13.7b, is 18 in less 2 in, or 16 in, and its width is, of course, 12 in:

$$M_u = \frac{w_u L^2}{2}$$

$$= \frac{0.20}{0.0020(6.5)}$$

$$= 15.3 \quad \text{or } 15 \text{ in}$$

This is a one-way slab situation, and using Table B.2(40/3), we can see that a moment of 2.8 kip-ft requires a ρ of only 0.0020. Trying #4 bars we get

$$s = \frac{a_s}{\rho d} \qquad (11\text{-}2)$$

$$= \frac{3.2(1.33)^2}{2}$$

$$= 2.8 \text{ kip-ft}$$

The longitudinal steel will be that required for temperature and shrinkage; i.e.,

$$A_s = \rho_t bh \qquad (11\text{-}1)$$

$$= 0.0020(36)(10)$$

$$= 0.72 \text{ in}^2$$

This can be provided by four #4 bars.

PROBLEMS

13.1 Evaluate the adequacy of a 7-ft-square footing for a column of 15 × 15-in section carrying a service live load of 70 kips and a service dead load (including the weight of the column) of 95 kips. The bottom of the footing is 5 ft below the surface of the soil, which weighs 115 pcf and can safely bear 4000 psf. The footing is 16 in thick with an effective depth of 11.75 in and is reinforced for moment with seven #6 bars of 60-ksi steel in each direction. Let $f_c' = 3500$ psi.

13.2 The footing thickness of Example 13.1 has been tentatively reduced to 16 in. Will the footing still be adequate in punching shear?

13.3 Design the footing for a masonry wall that is 15 in thick and carries service loads (including self-weight) of 8 kips/ft (dead) and 7 kips/ft (live). The bottom of the footing should be at a depth of 5 ft. The soil weighs 100 pcf and its allowable bearing capacity is 4000 psf. Use 3000-psi concrete and 60 000-psi steel.

13.4 A column footing adjacent to a property line has rectangular dimensions of 7 × 13.5 ft. It is 18 in thick, has an effective depth of 13 in, and is reinforced with six #9 bars in each direction; $f'_c = 3000$ psi and $f_y = 60$ ksi. The bottom of the footing is 4 ft deep in soil that weighs 100 pcf and has an allowable bearing capacity of 3500 psf. If the 16 × 16-in column carries a service dead load of 85 kips, which includes self-weight, and a service live load of 90 kips, is the design of this footing adequate?

13.5 Design the footing for an 8-ft-high concrete wall, 9 in thick, that carries a live load of 3 kips/ft and an applied dead load of 2.5 kips/ft. There is a 6-in-thick slab-on-grade carrying a live load of 50 psf. The bottom of the footing should be 3.5 ft below the surface soil, which weighs 110 pcf and has an allowable bearing capacity of 2500 psf. Let $f'_c = 3500$ psi and $f_y = 40\,000$ psi.

14

RETAINING WALLS

14.1 INTRODUCTION

There are three basic types of retaining walls as illustrated in Figure 14.1. The *gravity* wall is used in building construction only when a relatively small height is required, say, up to about 4 ft. Above 4 ft it becomes uneconomical in terms of the amount of concrete and the space required. The *cantilever* wall is the most frequently used type of retaining wall, effective up to about 20 or 25 ft in height. Above this height the *counterfort* wall becomes more efficient. The counterfort wall behaves like a one-way slab spanning horizontally to tapered beams which cantilever upward from a wide base.

FIGURE 14.1   Types of retaining walls.

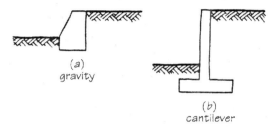

(a) gravity

(b) cantilever

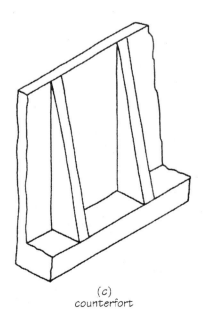

(c) counterfort

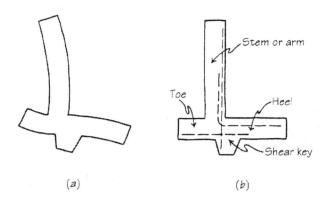

FIGURE 14.2 Cantilever retaining wall.

This chapter is concerned with the design considerations of *cantilever* walls. Figure 14.2*a* shows (in greatly exaggerated fashion) the deformations involved in a typical wall and Figure 14.2*b* shows the corresponding reinforcing patterns. These walls are usually designed such that the three kinds of failure shown in Figure 14.3 are prevented.

FIGURE 14.3 Failure modes.

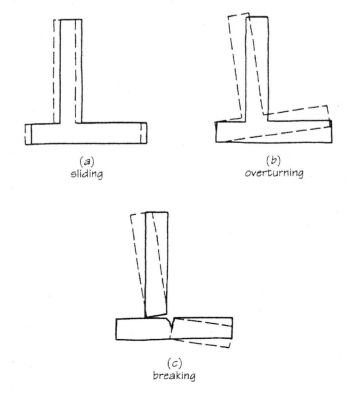

When the frictional forces that develop between the soil and the base of the wall are insufficient to prevent sliding, a shear key is added (Figure 14.2*b*). Most codes have traditionally specified a minimum factor of safety of 1.5 against both sliding and overturning. (Current trends indicate that a value closer to 2.0 is better for overturning.) Not only must the wall be safe against overturning about the front of the toe, but the bearing pressure on the soil (which maximizes at that point) must be kept below the safe bearing capacity. The pressure distribution under the base depends upon the location of the resultant of the sum of the vertical forces and the horizontal force as it passes through the base as shown in Figure 14.4. To keep the pressures low, it is desirable to keep the resultant within the middle third of the base dimension B.

The equations given in Figure 14.5 for the peak pressure at the toe of the wall were obtained by combining axial stresses with bending stresses due to the eccentric loading.

Other factors that influence design are shown in Figure 14.6. Each of these conditions serves to increase the lateral forces acting on the wall, and it is obviously better if such conditions can be avoided. The presence of water behind a wall is especially detrimen-

FIGURE 14.4

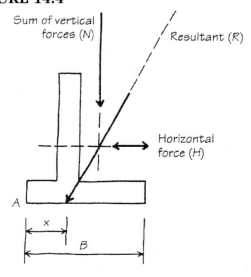

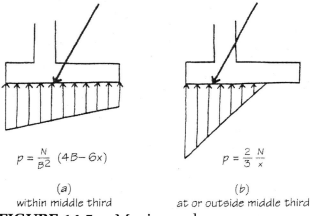

$$p = \frac{N}{B^2}\ (4B - 6x)$$

(a)
within middle third

$$p = \frac{2}{3}\ \frac{N}{x}$$

(b)
at or outside middle third

FIGURE 14.5 Maximum base pressures.

tal, and most designs include special drainage provisions (through the wall or behind it) to prevent the development of a pressure head. These special loading conditions will not be treated in this text.

It has been observed that many walls settle with time and rotate forward slightly. For this reason most designers give the wall a slight *batter* (slope from the vertical) to alleviate any unsafe appearances. Unless it is associated with design error or an overload condition, this slight rotation is not unsafe. It should be recognized that unlike the deformations that occur in most structural elements, this rotation actually serves to reduce the load on the wall.

14.2 LATERAL PRESSURES

The amount of horizontal pressure exerted by soils varies widely. A soft, moist clay will act very much like a fluid, whereas coarse, dry gravels will behave much less so. Classical theory indicates that the variation is related to the *internal angle of friction* ϕ of the soil, which can be determined by laboratory tests. (Strictly speaking, the ϕ angle is applicable only to cohesionless soils, but the concept is so widely used that adjustments are made so that reasonable ϕ values are obtainable for cohesive soils as well.)

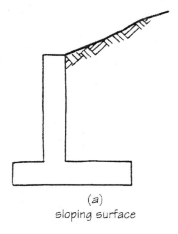

(a)
sloping surface

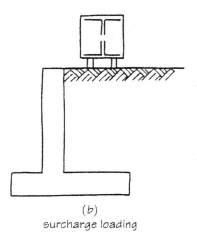

(b)
surcharge loading

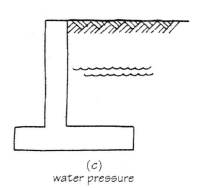

(c)
water pressure

FIGURE 14.6 Special loading conditions.

There are two kinds of lateral pressure that develop, *active* and *passive*, with the active pressure being much smaller for a given depth. The difference can be explained using the *sliding wedge* concept of Figure 14.7. In Figure 14.7a the soil is pressing against the

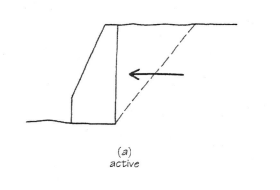

(a)
active

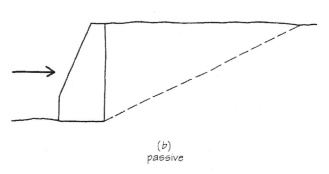

(b)
passive

FIGURE 14.7 Active and passive sliding wedges.

wall, and if the wall moves or deforms, the soil will eventually fail along the dotted line. Conversely, in Figure 14.7b the wall is being pushed into the earth and the failure plane that ensues will be much longer and take considerably more pressure to develop. (Passive pressures become very important in the design of a shear key at the base of a wall.) There is a reciprocal relationship between the active and passive pressure coefficients, given as

$$C_a = \frac{1 - \sin\phi}{1 + \sin\phi} \qquad (14\text{-}1a)$$

$$C_p = \frac{1 + \sin\phi}{1 - \sin\phi} \qquad (14\text{-}1b)$$

The coefficients are used to determine the lateral soil pressure assuming a linear varia-

tion with depth and a flat land surface with no surcharge:

$$p_a = C_a wh \qquad (14\text{-}2a)$$

$$p_p = C_p wh \qquad (14\text{-}2b)$$

where p_a and p_p = the active and passive lateral pressures, respectively

w = the unit weight of the soil

h = the depth below the surface

(The reader is cautioned that this approach to obtaining soil pressures is overly simplistic and should only be used for preliminary design purposes.)

14.3 PRELIMINARY DESIGN

Obtaining the proper proportions to meet the design requirements is usually done by selecting trial dimensions and performing an analysis on a one-foot strip of wall to see if the requirements can be satisfied. In the following example only the stability issues, soil pressures, and the adequacy of the stem in bending are considered. A complete design including the proper investigation of all parts for shear and moment and the establishment of various reinforcing patterns required by Code for such walls is beyond the preliminary design emphasis of this text.

Example 14.1 For the wall shown in Figure 14.8, make the following checks using f'_c = 3000 psi, f_y = 60 ksi, w_{soil} = 115 pcf, ϕ = 30°, f_r = 0.45, allowable bearing capacity = 3000 psf:

1. Safety against sliding (F.S. ≥ 1.5)

2. Safety against overturning (F.S. ≥ 2.0)

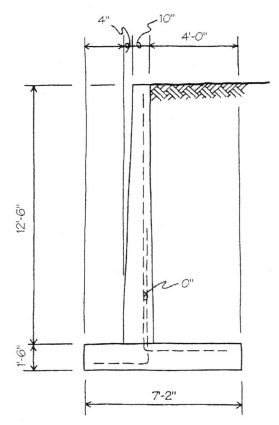

FIGURE 14.8 Wall for Example 14.1.

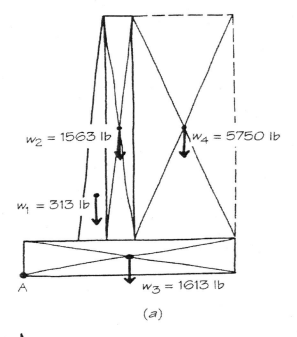

(a)

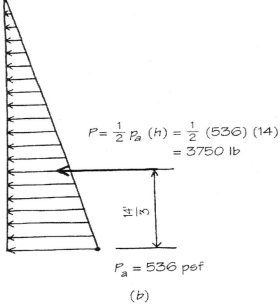

$$P = \frac{1}{2} P_a (h) = \frac{1}{2} (536) (14)$$
$$= 3750 \text{ lb}$$

$P_a = 536 \text{ psf}$

(b)

FIGURE 14.9 Forces on the wall.

3. Soil pressure less than allowable

4. Stem safe against breaking off at the base using a load factor of 1.6 and #5 bars placed vertically 8 in apart.

Solution: The forces acting on a one-foot strip of wall are shown in Figure 14.9. The active pressure is obtained by first getting the coefficient:

$$C_a = \frac{1 - \sin\phi}{1 + \sin\phi} \qquad (14\text{-}1a)$$

$$= \frac{1 - 0.5}{1 + 0.5}$$

$$= 0.333$$

Then

$$p_a = C_a wh \qquad (14\text{-}2a)$$

$$= 0.333(115)(14)$$

$$= 536 \text{ psf}$$

As shown in Figure 14.9, the lateral force resultant on the wall will then be 3750 lb.

The values of the different vertical forces are obtained using the given soil weight of 115 pcf and a concrete weight of 150 pcf. The sum of these weights is 9240 lb.

1. Without a shear key the only resistance to sliding is provided by friction between the soil and the bottom of the wall. It is customary to ignore the passive pressure resistance to sliding that might develop in front of the toe because the soil here can be fairly easily scoured away, at least for shallow depths such as those in this example. If the sum of the W forces is N, the frictional force is

$$F_r = f_r N$$
$$= 0.45(9240)$$
$$= 4160 \text{ lb}$$

This is clearly not enough to attain a factor of safety of 1.5, so a key will have to be added to this wall. The required depth h_k of the key can be determined with the aid of Figure 14.10. The passive pressure coefficient is the reciprocal of the active one; in this case, 3.0. The pressure at the base of the key (assuming scouring has removed the soil from in front of the toe) will be

$$p = 3.0(115)(h_k)$$
$$= 345 h_k$$

The resisting force provided by the key then will be

$$P_k = \frac{1}{2}(345) h_k^2$$

This force P_k will have to make up the difference between 1.5 times the sliding force P and the existing frictional force F_r:

$$P_k \geq 1.5(3750) - 4160$$
$$\geq 1465 \text{ lb}$$

Therefore, the minimum depth of key needed can be established by setting and solving for h_k:

$$h_k = 2.9 \text{ ft}$$

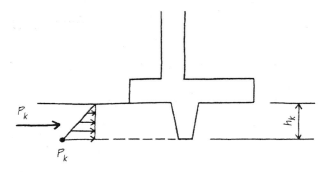

FIGURE 14.10   Passive pressure on the key.

[The reader is reminded that this would be a conservative value because the friction failure plane in front of the key would actually follow the dotted line in Figure 14.10 and the coefficient of friction between two soil masses is almost always greater than the soil-to-concrete coefficient. Similarly, the depth of the key is even more conservative if (as is the case in many designs) the toe could be placed below the ground surface developing a larger passive force.]

2. The safety against overturning can be determined with the help of Table 14.2. The moment arm values are those with respect to point A in Figure 14.9. The overturning moment is obtained by multiplying the sliding force P times its moment arm with respect to point A:

$$M_o = 3750\left(\frac{14}{3}\right) = 17\,500 \text{ lb-ft}$$

The factor of safety against overturning will be greater than 2.0:

$$\text{F.S.} = \frac{M_r}{M_o}$$
$$= \frac{40\,490}{17\,500}$$
$$= 2.3$$

(The actual F.S. will be slightly larger due to the weight of the shear key.)

3. To determine the maximum soil pressure under the toe, we need to establish where the resultant force intersects the base of the wall. Using statics and knowing from Varignon's theorem that the moment about point A of the N and H forces in Figure 14.4 is the same as the moment of their resultant R, we get distance x from point A:

$$x = \frac{M_r - M_o}{N}$$

$$= \frac{40\ 490 - 17\ 500}{9240}$$

$$= 2.49 \text{ ft}$$

This means that the resultant falls within the middle third of the wall and the peak pressure in Figure 14.11 will be obtained by using the equation from Figure 14.5a:

$$p = \frac{N}{B^2}(4B - 6x)$$

$$= \frac{9240}{(7.17)^2}\left[4(7.17) - 6(2.49)\right]$$

$$= 2470 \text{ psf}$$

This is less than the given soil bearing capacity of 3000 psf.

TABLE 14.2   Resisting moment

Vertical Force (lb)	Moment Arm (ft)	Moment (lb-ft)
$W_1 = $ 313	2.22	695
$W_2 = $ 1563	2.75	4 300
$W_3 = $ 1613	3.58	5 770
$W_4 = $ 5750	5.17	29 730
$N = 9240$ lb		$M_r = 40\ 490$ lb-ft

FIGURE 14.12

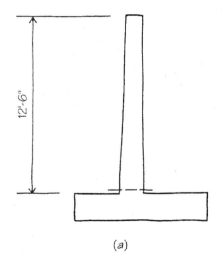

(a)

FIGURE 14.11

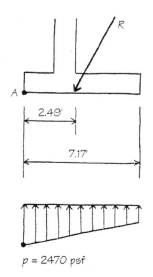

$p = 2470$ psf

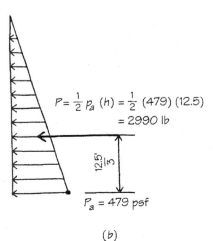

$P = \frac{1}{2}P_a\ (h) = \frac{1}{2}\ (479)\ (12.5)$

$= 2990$ lb

$P_a = 479$ psf

(b)

4. To establish the moment acting at the bottom of the stem, we need to find the lateral pressure at this location:

$$p_a = C_a wh$$

$$= 0.333(115)(12.5)$$

$$= 479 \text{ psf}$$

The total sliding force P is computed in Figure 14.12 as 2990 lb, and using a load factor of 1.6 because the ACI Code requires us to use the same factor as the one used for live load, we get

$$M_u = 1.6(2.99)\left(\frac{12.5}{3}\right)$$

$$= 19.9 \text{ kip-ft}$$

From Section 2.9, the proper cover requirement is 2 in clear and for a wall that is 1'-2" thick this means an effective depth of about 11.5 in. To be safe M_r must be at least as large as M_u.

Knowing that the stem is reinforced for moment by #5 bars spaced at 8 in, we can determine ρ.

$$s = \frac{a_s}{\rho d} \qquad (11\text{-}2)$$

or

$$\rho = \frac{a_s}{sd}$$

$$\rho = \frac{0.31}{8(11.5)}$$

The stem acts like a one-way slab, so for a 12 in length of wall, we can use Table B.2 (60/3). Interpolating between d values of 11 and 12 in, we find that

$$M_r = 24.0 \text{ kip-ft}$$

Since M_r is greater than M_u, the wall is safe in bending.

This is convenient because at some point up the wall alternate bars can be terminated leaving #5s at 16 in o.c. to carry the remaining moment. The Code stipulates that, just as for one-way slabs, the maximum bar spacing is three times the wall thickness, or 18 in, whichever is smaller.

The Code also requires horizontal reinforcing in the back of the stem as well as some steel in both directions for the front face. Besides having to resist bending in both directions, it is known that retaining walls are subject to unusual thermal stresses because one surface is exposed and the other is not.

PROBLEMS

14.1 The gravelly soil to be retained by the wall shown in Figure 14.13 has an internal angle of friction of 34° and a coefficient of sliding friction of 0.4; soil weight is 100 pcf. Considering 1.5 and 2.0, respectively, as minimum

FIGURE 14.13

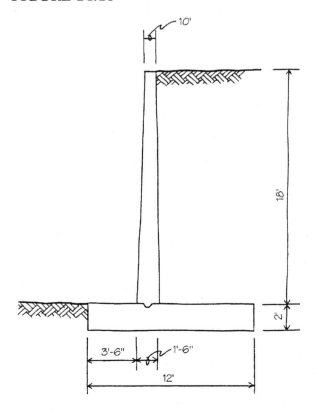

factors of safety, determine if the wall will be safe.
(a) Against sliding (if not, determine the h_k required)
(b) Against overturning
Will the 3500-psf allowable bearing capacity of the soil be exceeded?

14.2 Assuming an effective depth of 15.5 in at the section of maximum moment and using 60-ksi #6 reinforcing bars, determine reasonable spacing for moment steel at the base of the stem of the wall of Problem 14.1. Let f'_c = 3000 psi.

14.3 Determine the required depth of a shear key for the wall of problem 14.1 if the soil is composed mostly of clay and has an internal angle of friction of 30° and a coefficient of sliding friction of 0.3.

15

INTRODUC- TION TO PRESTRESSED CONCRETE

15.1 DEFINITIONS AND BASIC PRINCIPLES

The fundamental idea behind prestressing is presented in Section 10.5, in which it is shown that if an axial force is placed on a concrete column, the column will be able to accept more bending. This is true because the axial load causes all the fibers of the cross section to go into compression. If a bending moment is then applied, the column will develop stresses in both compression and tension; the compressive stresses will be algebraically additive to those already existing but the tensile stresses will be subtractive. Using appropriate load magnitudes, tensile stresses due to bending can be completely negated by the compressive stresses caused by the axial load. A fully loaded cross section with zero tensile stress obviously has desirable implications for a reinforced concrete beam.

Prestressing may be defined as the purposeful introduction of stresses into a member so that undesirable stresses resulting from the applied loads may be reduced.

While the idea of prestressing concrete was explored much earlier, the most significant developments took place in Europe in the 1930s and were often pioneered by Eugene Freysinnet (1879-1962), a French engineer.

Prestressed concrete is usually more expensive than regular reinforced concrete, but it has a number of distinct advantages. It can span farther with less depth; tension cracks can be reduced or eliminated; and beam deflection can be reduced or eliminated. The smaller cross sections also mean less structural dead weight, always a concern with concrete structures.

The higher cost is due, in part, to the fact that higher-strength materials are required to obtain the benefits of prestressing. The prestress force is usually imparted to the concrete by embedded steel strands which are stretched elastically by hydraulic jacks. When the jacking force is removed, each steel strand attempts to return to its original length; it is prevented from doing so by the concrete, which goes into compression. Ordinary steel cannot be used for prestressing for the simple reason that it will not strain very much before yielding. If a reinforcing bar with a yield strength of 60 ksi is stretched so that it has a tensile stress of 30 ksi, its strain level will be about 0.0010. This is about equal to the strain that will result in the concrete due to shrinkage and creep. In other words, in time the concrete will shorten about as much as we had stretched the steel and no prestress will be left. If, however, high-strength cable steel is used, with a yield value of 200 ksi or greater, and it is stressed to 100 ksi, plenty of prestress will still be left after the concrete shortens. The high prestress forces needed to make the steel effective also result in the need for higher-strength concrete, and f'_c values of 6 and 8 ksi are not uncommon. Depending upon the application, the higher cost of these materials can be more than offset by the advantages of prestressing.

There are essentially two kinds of prestressing used with concrete: pretensioning and posttensioning. The prefix relates to the time at which the steel is stretched (i.e., before or after the concrete is placed). In *pretensioning*, the technique usually used with precast beams, many steel strands about $\frac{7}{16}$ or $\frac{1}{2}$ in in diameter are stretched between two rigid steel abutments on a prestressing bed. The most commonly used strand is composed of seven wires, six of which spiral around a center wire.

The distance between the abutments may be 300 ft or more as several members placed end to end can be made at one time with this technique. After the strands are all stretched to the desired tension and held by the abutments, the concrete is placed in the steel forms and cured to the minimum strength necessary to accept the prestress force. The strands are then cut between each of the

FIGURE 15.1 Posttensioning tendons.

members and the prestress force is imparted to the concrete by bond. The members are removed from the forms for further curing, enabling the forms to be reused immediately. Pretensioning has the high level of quality control normally associated with factory production environments.

Posttensioning can be used with cast-in-place concrete and is done by placing the concrete around one or more tendons that have been positioned in the form. These tendons, often about 2 in in diameter, are normally draped as shown in Figure 15.1 and are coated or sheathed with some plastic material to prevent bonding with the concrete. Alternatively, they are placed in conduits or passageways left in the concrete. After the concrete has been cured to the desired strength level, the tendons are stretched and fixed to anchorage plates at their ends. The prestress force is imparted to the concrete via these end anchorages. The lack of bond between the tendons and concrete means that tensioning can occur in stages if required by the design. (It is sometimes desirable to introduce further prestress force after some of the dead loads have been applied to the beam.)

Whether pretensioned or posttensioned, all prestressed members have other reinforcing as well, sometimes for flexure and usually for shear. Pretensioned members must be reinforced for handling and posttensioned members reinforced for high localized anchorage stresses.

Because posttensioning is done on site, it is generally used with regular monolithic concrete construction. After construction is complete it is usually not possible to tell whether a beam is posttensioned or simply reinforced, except perhaps by its high span/

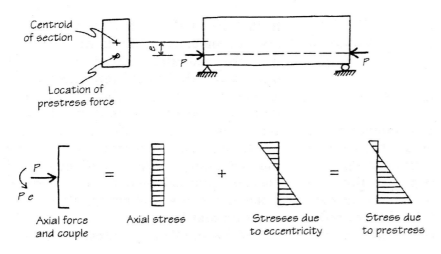

FIGURE 15.2 Eccentric prestress force.

depth ratio. The nature of the building is still characteristically concrete and generally has the continuity achievable with this material. Precast and pretensioned construction, on the other hand, assumes a different nature. The architecture that results is often one which reflects the fact that the building is made from prefabricated pieces or parts. In this respect the design vocabulary is similar to that of wood and steel and less like concrete in the usual sense. In precast construction the real structural design issues are often associated with how the pieces join rather than how they affect one another in terms of moments and deflections. These are general observations, of course, and are subject to exception. For instance, posttensioning techniques are sometimes used to "fasten" precast segments of buildings together at the site, and continuity is achieved in a different fashion.

The location of the prestressing steel is very important in both pretensioning and posttensioning. With a concentric prestress force, the distribution of the axial stresses is uniform over the cross section; this means that the compressive stress will always be increased by the same amount that the tensile stress is reduced. A large prestress force in combination with bending stresses can result in an overstress of the concrete in compression. This can be avoided by moving the prestress force away from the centroid toward the tension side of the beam. It is then called an *eccentric prestress force*, and its effects are shown in Figure 15.2.

An eccentric force can be treated as having two components, an axial force and a couple. By changing the eccentricity e, we can manipulate the prestress distribution. A prestress force placed below the centroid will result in a slight upward camber in the beam (depending upon the magnitude of the self-weight), and it is usually considered desirable to induce a little tension in what will become the extreme compressive fiber.

In the dual signs of the following equations the top sign refers to the top fiber stress and the bottom sign refers to the bottom fiber stress. The two effects of an eccentric prestress force can be computed as

$$f_p = -\frac{P}{A} \pm \frac{Pec}{I} \qquad (15\text{-}1)$$

where the values f_p are the top and bottom stresses due to prestress. As usual, the positive sign denotes tension and the negative sign denotes compression.

The addition of the regular bending stresses f_b from the live and dead loads is shown in Figure 15.3. Once again, by proper location of the prestress force the final stresses can be manipulated. It is even possible to have a section completely in compression

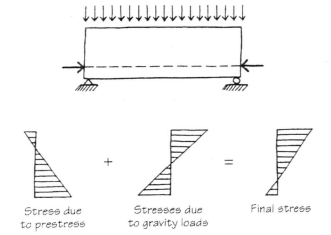

Stress due to prestress   +   Stresses due to gravity loads   =   Final stress

FIGURE 15.3   Prestresses plus bending stresses.

under all the loads. Since $f_b = Mc/I$, the final stresses are obtained as follows:

$$f = -\frac{P}{A} \pm \frac{Pec}{I} \mp \frac{Mc}{I} \qquad (15\text{-}2)$$

where f represents the final top and bottom stresses.

Obviously, it would seem ideal to be able to vary the location of the prestress force over the length of the beam in response to a changing gravity load moment. This is relatively easy to do with the draped tendons used in posttensioning, and continuous beams are often prestressed using continuous tendons such as that of Figure 15.4. Even the necessarily straight tendons used in pretensioning are often placed on a slope in elevation so that the maximum eccentricity occurs at the point of greatest moment.

Prestressed concrete is used for floor slabs as well as for beams. Many flat plate systems are constructed with posttensioning tendons running in both directions. They are spaced

FIGURE 15.4   Varying tendon eccentricity (shown greatly exaggerated).

over the entire floor and are often continuous from one side of the building to the other, moving up and down within the slab as required by the moments. For buildings having a relatively constant structural bay size with no unusual loads or conditions the use of precast pretensioned planks can often result in considerable savings. Under light floor or roof loads a highly prestressed plank system can span up to about 40 ft.

Most manufacturers of pretensioned concrete have standard planks and other sections, such as the typical ones shown in Figure 15.5. In many cases they can be selected directly from load tables provided by the manufacturer. Because of the cost of transportation, such precast sections are usually more popular for building sites that are not great distances from the precasting plant.

15.2   BASIC ANALYSIS OF PRESTRESSED BEAMS

The complete analysis and design of prestressed members calls for the use of both allowable stress and strength design procedures. Unlike regular reinforced concrete, the cracks on the tension side of a prestressed member are infrequent, if not absent altogether. Most beams have compression over the entire cross section except at locations of maximum moment, and even at these points tensile stresses develop only under the full service load. In the absence of cracks prestressed concrete behaves much more like a homogeneous material than like reinforced concrete. The basic flexure theory has more validity, and many structural designers use allowable or working stress techniques to proportion members and strength methods as a final check on their capacities. In the brief introduction provided in this section and the one that follows we use only working stress methods and check only a few stresses. A

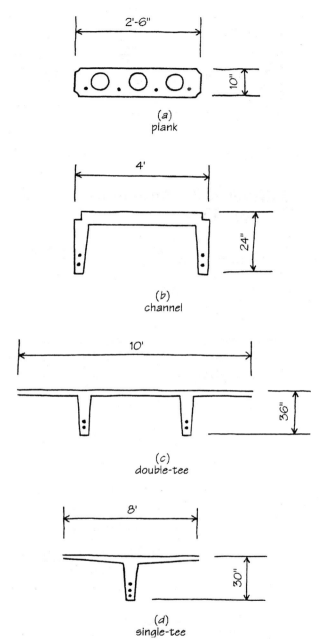

(a)
plank

(b)
channel

(c)
double-tee

(d)
single-tee

FIGURE 15.5 Typical standard precast sections with representative dimensions.

thorough analysis of prestressed concrete behavior is beyond the intent of this text and can be found in numerous books that are devoted entirely to the subject. (Consequently, the examples and problems in this chapter have been somewhat simplified and not all factors which need to be considered have been included.)

Example 15.1 The prestressed beam in Figure 15.6 has two tendons for a total area of 4 in^2. The stress in the tendons is 90 ksi. Assuming that the uniform load shown *includes* the beam selfweight, find the top and bottom fiber stresses at midspan.

Solution: First, find the stresses due to the prestress force alone:

$$P = f_y A_s$$
$$= 90 \text{ kips/in}^2 (4 \text{ in}^2)$$
$$= 360 \text{ kips}$$

From the previous section,

$$f_p = -\frac{P}{A} \pm \frac{Pec}{I} \tag{15-1}$$

FIGURE 15.6 Prestressed beam

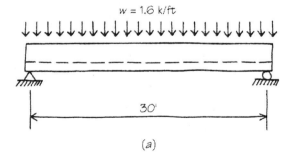

$w = 1.6$ k/ft

30'

(a)

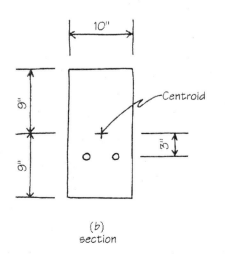

10"

9"

9"

3"

Centroid

(b)
section

For this cross section we can find that $A = 180 \text{ in}^2$ and $I = 4860 \text{ in}^4$:

$$f_p = -\frac{360 \text{ kips}}{180 \text{ in}^2} \pm \frac{360 \text{ kips}(3 \text{ in})(9 \text{ in})}{4860 \text{ in}^4}$$

$$f_{p_{top}} = -2 \text{ ksi} + 2 \text{ ksi} = 0$$

$$f_{p_{bottom}} = -2 \text{ ksi} - 2 \text{ ksi} = -4 \text{ ksi}$$

These prestresses are shown in Figure 15.7a and are constant over the full length of the beam.

The midspan bending stresses are given by

$$f_b = \mp \frac{Mc}{I}$$

$$M = \frac{wL^2}{8}$$

$$= \frac{1.6 \text{ kips/ft}(30 \text{ ft})^2}{8}$$

$$= 180 \text{ kip-ft}$$

$$f_b = \mp \frac{180 \text{ kip-ft}(12 \text{ in/ft})(9 \text{ in})}{4860 \text{ in}^4}$$

$$f_{b_{top}} = -4 \text{ ksi}$$

$$f_{b_{bottom}} = +4 \text{ ksi}$$

These stresses are shown in Figure 15.7b, and the combined stress distribution appears in Figure 15.7c. Notice that, owing to the prestress, the entire section is in compression under the full service load.

15.3 INITIAL AND FINAL STRESSES

The basis for any type of working stress or service load design is that the allowable stresses for the material(s) never be exceeded, i.e., the factor of safety inherent in these stresses must not be reduced under the worst possible anticipated loading condition. It is important to recognize that with prestressed

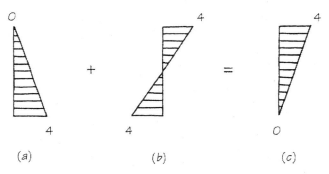

FIGURE 15.7   Stress distributions for Example 15.1.

concrete at least two stages may be critical. The first occurs when the prestress is initially applied to the concrete. With pretensioning, this is referred to as the time of transfer; with posttensioning, it happens when the tendons are "jacked." Usually at this time the concrete is not fully cured and may be quite weak. The economics of construction do not permit the lapse of several weeks before stressing the concrete. With eccentrically located tendons, tension usually exists on the far edge of member and compression on the edge nearest the tendons. At this stage the self-weight of the member is usually the only load that acts to reduce these stresses, so this condition can become critical. It must also be remembered that the prestress values usually vary over the length of beam as do the self-weight stresses.

The second and more obvious stage occurs when the beam is finally loaded by the full service load. Now the concrete has presumably reached its full (28-day) strength. The load acts in opposition to the prestress, and what was the tensile fiber before goes into compression and vice versa. These final stresses must not exceed the allowable values.

The American Concrete Institute (ACI) has provided two sets of allowable stresses for use with prestressed members, one for the initial application of the prestressing force and a second for the final loading stage. These are summarized in Table 15.1. The quantity f'_{ci} refers to the cylinder strength of the concrete at the time of prestress.

TABLE 15.1 *ACI allowable stresses for prestress concrete*

Stage	Allowable Stresses (psi)	
	Tension	Compression
Initial (at prestress, except at the ends of a simple beam)	$+3\sqrt{f'_{ci}}$	$-0.60f'_{ci}$
Initial (at prestress at the ends of a simple beam)	$+6\sqrt{f'_{ci}}$	$-0.60f'_{ci}$
Final (at service load)	$+6\sqrt{f'_c}$	$-0.45f'_c$

Note: The values of f'_{ci} and f'_c under the radical sign must be expressed in units of psi.

This analysis at more than one stage of loading is further complicated by the fact that the prestress force is not constant. It drops considerably between the initial and final stages, with the difference called the *prestress loss*. The amount of this loss depends upon many factors, including shrinkage, creep, elastic shortening, and friction, and is generally about 15 or 20%. It is clear that different values of the force must be used for the different stress checks. The initial prestress force is usually denoted as P_i and the final force as P when using Equations (15-1) and (15-2).

The ACI Code recommends allowable steel stresses that are quite high, 70 or 80% of the ultimate stress, for two reasons. First, the steel stress will always drop from its initial value, and second, the stress is known with much more than the usual order of construction accuracy. For one thing, it is very easy to measure the steel elongation during the jacking operation and use the modulus of elasticity to get the stress in a tendon.

Example 15.2 The beam section in Figure 15.8 is pretensioned with straight strands, stressed to their maximum allowable value of 150 ksi. The span is 24 ft. Use the allowable stresses in Table 15.1.

1. Determine the maximum permitted area of prestressing steel.

2. If the prestress losses are 20%, find the maximum uniform load that can safely be applied to this beam.

Solution: The area of the. cross section is 168 in^2 and the moment of inertia computes to 5824 in^4. The eccentricity is given as 4 in.

1. For this condition it can be shown that the amount of steel will be controlled by the initial allowable stresses at a section near the end of the beam where the prestress is not tempered by the self-weight. It can be controlled by either the tensile stress or the compressive stress. The extreme fiber stresses at the time of prestress are measured by

$$f_p = -\frac{P_i}{A} \pm \frac{P_i ec}{I} \qquad (15\text{-}1)$$

From Table 15.1, the allowable tensile stress is $6\sqrt{f'_{ci}}$ and the allowable compressive stress is $0.60\, f'_{ci}$. In this case the needed values are, respectively, 355 psi (0.355 ksi) and 2100 psi (2.1 ksi). If the tensile stress at the top governs, then by

FIGURE 15.8 Inverted T beam.

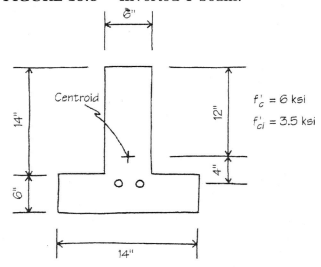

substituting +0.355 ksi into Equation (15-1), we can find the limiting value of the initial prestress force:

$$+0.355 \text{ ksi} = -\frac{P_i}{168 \text{ in}^2} + \frac{P_i(4 \text{ in})(12 \text{ in})}{5824 \text{ in}^4}$$

Solving for P_i gives us 155 kips.

To determine if the compressive stress at the bottom governs, we use −2.1 ksi in the same equation to get a different value of P_i. (Note the signs!)

$$-2.1 \text{ ksi} = -\frac{P_i}{168 \text{ in}^2} - \frac{P_i(4 \text{ in})(8 \text{ in})}{5824 \text{ in}^4}$$

Solving for P_i this time, we get 183 kips.

Evidently, the tensile stress will control and the initial prestress force can be no larger than 155 kips. If the steel is stressed to 150 ksi as given, then its area can be no larger than

$$\frac{155 \text{ kips}}{150 \text{ ksi}} = 1.03 \text{ in}^2$$

2. After losses the prestressing force becomes 0.80 (20% loss was given) times 155 kips, or 124 kips.

ACI Code allowable stresses for the final state (when the concrete has reached its 28-day strength) are $6\sqrt{f'_{ci}}$ in tension and $0.45 f'_c$ in compression, or, in this case, 465 psi (0.465 ksi) and 2700 psi (2.7 ksi), respectively:

$$f = -\frac{P}{A} \pm \frac{Pec}{I} \mp \frac{Mc}{I} \qquad (15\text{-}2)$$

We can use Equation (15-2) to determine the largest permissible gravity load moment. If the bottom fiber tension controls, then

$$+0.465 \text{ ksi} = -\frac{124 \text{ kips}}{168 \text{ in}^2} - \frac{124 \text{ kips}(4 \text{ in})(8 \text{ in})}{5824 \text{ in}^4}$$
$$+ \frac{M(8 \text{ in})}{5824 \text{ in}^4}$$

Solving for M, we find

$$M = 1372 \text{ kip-in} \text{ or } 114 \text{ kip-ft}$$

If the top fiber compression controls, then

$$-2.7 \text{ ksi} = -\frac{124 \text{ kips}}{168 \text{ in}^2} + \frac{124 \text{ kips}(4 \text{ in})(12 \text{ in})}{5824 \text{ in}^4}$$
$$- \frac{M(12 \text{ in})}{5824 \text{ in}^4}$$

Solving for M gives us

$$M = 1448 \text{ kip-in} \text{ or } 121 \text{ kip-ft}$$

Again, the tensile stress controls. The maximum moment is 114 kip-ft and the maximum total load is readily computed for the simple beam as

$$w = \frac{8M}{L^2}$$
$$= \frac{8(114 \text{ kip-ft})}{(24 \text{ ft})^2}$$
$$= 1.58 \text{ kips/ft}$$

The self-weight of the beam per linear foot is

$$\frac{168}{144}(150) = 175 \text{ lb} \text{ or } 0.175 \text{ kips}$$

Therefore, the applied load that can be safely carried by the beam is 1.58 less 0.175, or approximately 1.4 kips/ft.

PROBLEMS

15.1 The pretensioned beam in Figure 15.9 has a group of six deflected tendons. The load w is 1 kip/ft, which includes the beam selfweight. Let the prestress force be 160 kips and determine the midspan extreme fiber stresses.

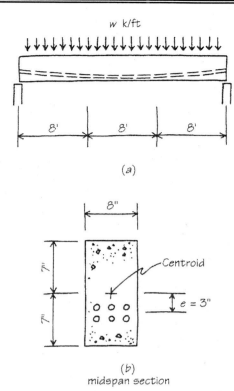

FIGURE 15.9   Prestressed beam.

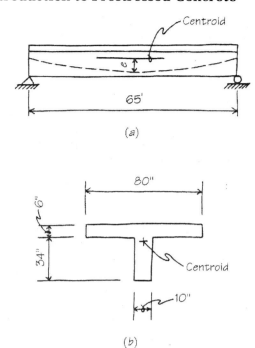

FIGURE 15.10   Posttensioned T beam.

15.2 The T beam of Figure 15.10 has a prestress force of 1350 kips. The maximum allowable tensile stress in the concrete used is 300 psi at the time of tensioning. If the only load on the beam is its own self-weight, compute the maximum permissible tendon eccentricity at midspan.

15.3 The T beam of Figure 15.10 has a prestress force of 1800 kips with a midspan eccentricity of 8 in. How much load w kips/ft, including the self-weight, can be placed on this

beam if the concrete stresses at the time of loading are limited to 3000 psi in compression and 580 psi in tension? Assume that the midspan section controls.

15.4 The roof plank in Figure 15.11 must carry a load of $w = 0.3$ kip/ft, which includes its own self-weight. Its design is based on ACI allowable stresses.
  (a) Determine the maximum permitted prestress force.
  (b) If the prestress loss is 18%, how far can the plank safely span under the given load?

FIGURE 15.11   Precast plank

$f_c' = 5800$ psi cross section.

$f_{ci}' = 2200$ psi

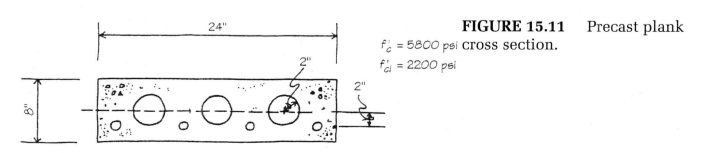

15.5 The cross section in Figure 15.12 is used for a simply supported girder spanning 65 ft. In addition to its own weight, the girder must carry an applied live load of 0.3 kip/ft and an applied dead load of 0.7 kip/ft. Assume that the design is governed by ACI allowable stresses and that the concrete has its full strength of 7250 psi at the time of jacking (i.e., $f_{ci}' = f_c' = 7250$ psi). Assume that the stresses at midspan will control.

(a) If P_i is 675 kips, determine the maximum permissible eccentricity if the load on the member is its own weight plus the dead load of 0.7 kip/ft.

(b) If the prestress loss is 20%, are the stresses within the allowables under the full load?

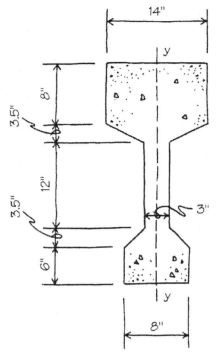

FIGURE 15.12 Prestressed concrete section.

16

PRELIMINARY STRUCTURAL DESIGN OF A BUILDING

16.1 DESCRIPTION AND ASSUMPTIONS

This book is concerned with preliminary structural design and the basic behavior of reinforced concrete buildings. This chapter illustrates how the principles and practices explained in previous chapters are applied to a simple one-way slab, three-story building. The preliminary design of a typical slab, beam, girder, column, and footing are presented, all with the goal of checking the feasibility of the structural scheme in terms of member sizes. Approximate loading patterns, moment coefficients, and moment estimates are used in lieu of an exact structural analysis, which would be needed for the final design. Only gravity loads are considered as lateral loading is not a subject of this text. (The building is not very tall and unless it is located in a high wind or seismic risk area, the forces due to lateral loads are not likely to control the design anyway.)

The much used provision found in most building codes called *live load reduction* is also omitted in these preliminary designs. This provision, alluded to briefly in Section 12.4, makes use of the statistical probability that a supposedly uniform code-specified live load will not be present on every square foot of a member's tributary area at the same time. This is particularly true for members with large tributary areas, such as girders, columns, and footings. Thus, the codes usually permit such elements to be designed for less than their full complement of live load.

The essential features of the building are illustrated in Figures 16.1, 16.2, and 16.3. It is a marketing center for the wholesale display and storage of large appliances. The two bay sizes were determined by architectural plan considerations involving the location of offices, corridors, and display spaces. The architect considered three structural deck systems:

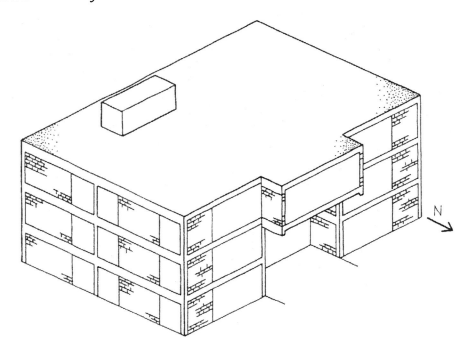

FIGURE 16.1   Isometric.

1. One-way slab with beams (the subject of this investigation)

2. One-way pan system

3. Two-way waffle pan system

The waffle slab was rejected on the basis of cost and the fact that it would be hidden for the most part by a suspended ceiling. (Had the spans been longer, it would have received further consideration.) The one-way pan system was a viable alternative and could easily have been selected. Indeed, if the live loads had been any larger, it probably would have been selected.

A first guess at a typical column size has been made at 16 × 16 in. In the interest of construction simplicity the beams and girders are assumed to have the same width as the columns.

Deflection will not be checked for the beams and girders as it seldom governs the design of a continuous bending member unless:

1. The member is wide and shallow.

2. The member is carrying mostly dead load and creep becomes an issue.

Similarly, shear seldom controls the overall dimensions of a beam unless such a beam is quite short and heavily loaded. Shear can be a problem, however, for any member carrying large concentrated loads. For this reason stirrup placement will be checked for the girder design as the girders receive most of their load via the concentrated beam reactions. In determining the loading patterns, the effects of openings in the slabs for vertical circulation and chase spaces will be ignored. This greatly simplifies the tributary area determinations and will have a relatively small effect on member sizes anyway. For similar reasons a live load value of 100 psf is assumed to act over the entire floor area even though the office areas require only a 60-psf live load, as seen in Table 16.1.

For the continuous slabs and beams we assume that the governing situation occurs when the full load values, live and dead, are applied to all spans. This will result in negative maximum moments. That is, we assume that the ratio of live to dead loads will

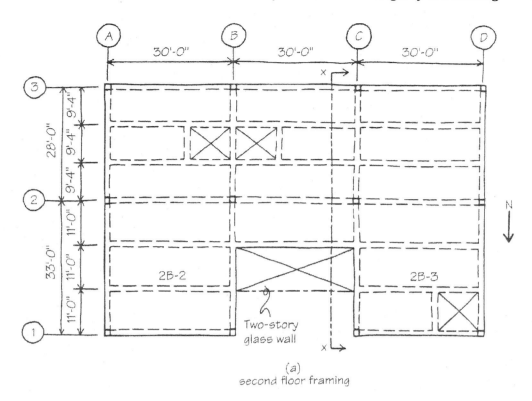

(a)
second floor framing

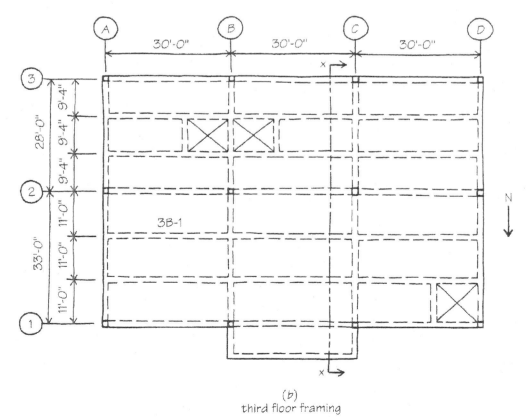

(b)
third floor framing

FIGURE 16.2

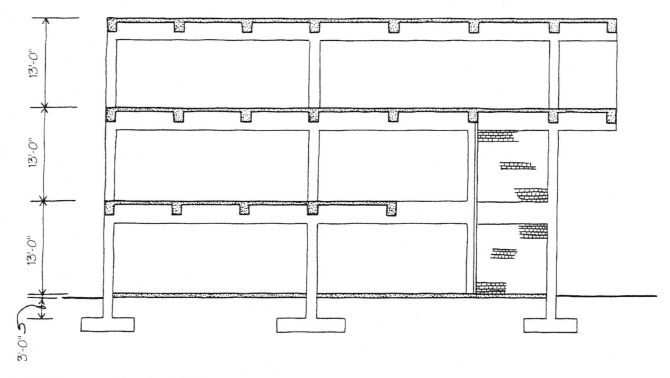

FIGURE 16.3 Section X-X.

not be so great as to cause the governing situation to occur when the live load is present on alternate spans only, thus creating large positive moments. (See Section 4.2.)

16.2 ONE-WAY SLAB DESIGN FOR A TYPICAL FLOOR

The slab spans for the 33-ft bays are slightly larger than those for the 28-ft bays, and assuming that a single slab thickness is used throughout, the larger span will determine the slab thickness. As seen in Figure 16.4, if the preliminary assumption of 16 in is used for the column and beam widths, a 9'-8"-slab span will result. From Table 11.1, the recommended minimum thickness will be 1/24 of the span if the torsional stiffness of the spandrel beam is ignored, which is conservative. Thus, the span of 116 in calls for a thickness of 5 in.

The continuous slab actually has six spans

and the amount of positive and negative moment steel varies depending upon the span dimension and support conditions. In lieu of an indeterminate analysis, the American Concrete Institute (ACI) coefficients of Figure 4.18b may be used. Since we have more than two spans, the largest moment will be a negative end span value of $wL^2/10$. The slab itself weighs

$$\frac{5}{12}(150) = 63 \text{ psf}$$

The total service dead load then will be 63 plus the 17 psf of miscellaneous dead load from Table 16.1. The full factored uniform load becomes

$$w_u = 1.2(63 + 17) + 1.6(100)$$

$$= 266 \text{ psf} \approx 0.27 \text{ ksf}$$

The design moment is then

$$\frac{w_u L^2}{10} = \frac{0.27(9.67)^2}{10} = 2.5 \text{ kip-ft}$$

TABLE 16.1   *Load and strength data*

Gravity Loads			Materials		
Roof			f'_c		
	DL (misc.)	= 10 psf	Footings		3500 psi
	LL	= 30 psf	Slabs, beams, and girders		4000 psi
			Columns		4000 psi
Floor					
	DL (misc.)	= 17 psf	f_y		
	(includes interior		Slab steel		40 ksi
	partitions)		All other steel		60 ksi
	LL				
	(display space)	= 100 psf	Soil bearing capacity		5500 psf
	(office space)	= 60 psf	Soil weight		100 pcf
	(corridors)	= 100 psf			
Walls					
	Glass	= 3 psf			
	Masonry	= 80 psf			

Maintaining the ¾-in cover requirements for slabs, if #4 bars are used, the effective depth needed will be 4 in. Table B.2 (40/4) provides the required steel ratio for the given material strengths of f'_c = 4000 psi and f_y = 40 ksi. This table shows that a ρ value of 0.0045 provides a moment capacity of exactly 2.5 kip-ft. We can find the corresponding bar spacing from Equation (11-2):

$$s = \frac{a_s}{\rho d} \qquad (11\text{-}2)$$

Since the area of a #4 bar is 0.20 in², we get

$$s = \frac{0.20}{0.0045(4)}$$
$$= 11 \text{ in}$$

This spacing is less than the Code maximum values of $3h$ and 18 in.

The required spacing for other locations in the slab can be determined in a similar fashion. By using the largest moment value for this spacing, we can establish the sufficiency of the 5-in slab thickness.

Equation (11-3) can be used to find the required spacing of temperature steel, which will be perpendicular to the moment steel. (It normally rests on top of the positive steel.) Grade 40 steel requires a ρ_t of at least 0.0020, and if #4 bars are used, we obtain

$$s = \frac{a_s}{\rho_t h} \qquad (11\text{-}3)$$
$$= \frac{0.20}{0.0020(5.0)}$$
$$= 20 \text{ in}$$

FIGURE 16.4   Section through a typical 33-ft bay.

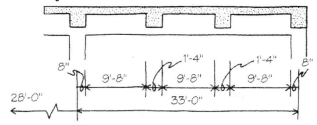

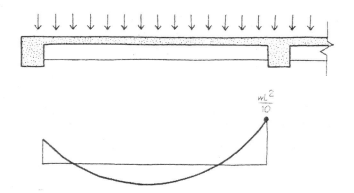

FIGURE 16.5 Estimated maximum moment in beam 3B-1.

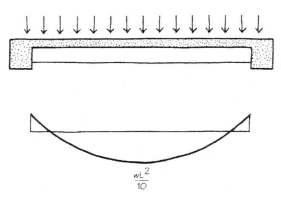

FIGURE 16.6 Estimated maximum moment in beam 2B-2.

This is larger than the code maximums of $5h$ and 18 in, so 18 in should be used. Alternatively, #3 bars can be used, which will call for an 11-in spacing. However, unless an uneconomical design results, construction is simplified if the same bar size is used throughout the slab.

16.3 BEAM DESIGN

For most of the building the east-west beams are continuous over three equal spans. From the ACI moment coefficients in Figure 4.18b, higher moment values will occur in an end span that frames into a spandrel girder as opposed to one that meets a column. The beam with the mark 3B-1 in Figure 16.2b represents this case, and the ACI coefficients indicate that a reasonable maximum moment value will be negative $wL^2/10$. This moment is shown in Figure 16.5.

It is also important to notice the single span beams on the second floor, which bear the marks 2B-2 and 2B-3 in Figure 16.2a. The building's two-story entry interrupts this beam line, and depending upon the torsional stiffness of the girders, 2B-2 and 2B-3 could approach a simply supported condition. In such a case the maximum moment would be positive with an upper limit of $wL^2/8$. Again, a reasonable design value might be $wL^2/10$ because the girders would provide at least a small amount of negative moment at the beam ends, which would reduce the positive moment somewhat. This is shown in Figure 16.6.

Figure 16.7 shows a section through a beam and slab portion in an area where the clear span is 9 ft-8 in. From the previous section, the ultimate load on the slab is 0.27 ksf, which includes the effect of the slab's own weight. The tributary strip length is 11.0 ft, so

FIGURE 16.7
Tributary strip for a beam.

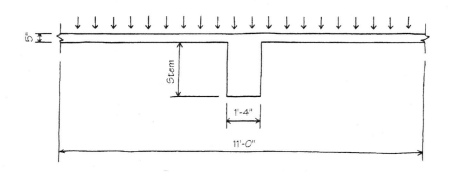

the total load on the beam from the slab will be

$$11.0 \times 0.27 \text{ ksf} = 3.0 \text{ kips/ft}$$

To this the factored weight of the beam stem must be added. The beam width is 16 in but the beam depth is as yet undetermined, so this weight must be estimated. Assuming a beam stem extension below the slab of 20 in, its weight becomes

$$\frac{16(20)}{144}(150) = 333 \text{ plf} \approx 00.33 \text{ klf}$$

Its factored weight will be

$$1.2(0.33) = 0.4 \text{ klf}$$

The ultimate design load including the stem self-weight will be

$$w_u = 3.0 + 0.4 = 3.4 \text{ klf}$$

(The self-weight of that portion of the beam within the slab was included in the slab load.)

For continuous beam analysis the code specifies the use of center-to-center support distance for span lengths and then permits reductions in the end moments for the effect of the support thickness and for the "redistribution of moments" that occur at or near failure loads. In this text we have ignored those reductions and used the center-to-center span length when studying frame equilibrium and the clear span distance in obtaining moments for preliminary design. From Figure 16.2, the clear span of all the beams will be 28 ft-8 in, obtained by subtracting two girder half-widths (a total of one girder width, 16 in) from the center-to-center of distance 30 ft-0 in. The design moment (negative for 3B-1 and positive for 2B-2) will be

$$M_u = \frac{w_u L^2}{10}$$

$$= \frac{3.4(28.7)^2}{10}$$

$$= 280 \text{ kip-ft}$$

(Note that this implies a level of accuracy that does not exist; if our estimating is on the conservative side, then the true value is probably some place in between, say, 250 and 300 kip-ft.)

For preliminary purposes we prefer to use a steel ratio of $\frac{2}{3}\rho_{\max}$. From Table A.2, $\rho_{\max}$ is 0.0213. Therefore,

$$\rho = \frac{2}{3}(0.0213)$$

$$= 0.0142$$

From Table B.1(4), the coefficient of resistance R will be 0.745. Using Equation 7-2b to find the effective depth needed to accommodate our M_u, we get

$$d = \sqrt{\frac{M_u}{\phi b R}} \qquad (7\text{-}2b)$$

$$= \sqrt{\frac{280(12)}{0.9(16)(0.745)}}$$

$$= 17.7 \text{ in} \approx 18 \text{ in}$$

[Note that Graph C.1 (4) can be used to get the same value less accurately.] The overall beam depth (thickness) will be about 22 or 23 in depending upon the bar sizes and how the steel is detailed. The clear cover requirement from Figure 2.9 is 1½ in, and it is customary to allow a ½-in space for the shear stirrups. It may also be noticed that where the beam intersects a girder at a column, the negative moment steel of the beam must cross over or under that of the girder. It is better to provide more clearance here rather than less. In any event Figure 16.8 confirms the sufficiency of the estimate of 20 in for the beam stem.

The amount of steel needed will be

$$A_s = \rho b d \qquad (6\text{-}7)$$

$$= 0.0142(16)(18)$$

$$= 4.03 \text{ in}^2$$

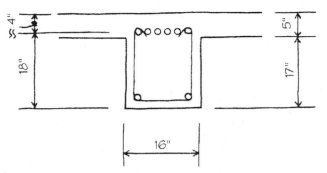

FIGURE 16.8   Steel at the end of 3B-1.

Table A.1 shows that six #8 bars will provide 4.72 in² and Table A.3 indicates that they will fit within our 16-in width. The same amount of steel will be required for the positive moment of beam 2B-2 and the overall size of either beam will be 16 × 22 in. The continuous beams in the 28-ft bay of the building will be smaller because they have less tributary area.

The spandrel beams along the north and south walls will carry less floor area than an interior beam but must carry the exterior wall. If this wall is heavy, the total load on the beam is likely to be about the same as for an interior beam. If the wall is light, then most of the load will come into the beam from one side creating torsional forces. For these reasons spandrel beams and girders often end up having the same overall dimensions as their interior counterparts.

As mentioned in Section 16.1, stirrup placement will not be considered in the preliminary design of the beams in this structure. This spacing of stirrups seldom controls the size of any concrete beam or girder, especially those carrying only uniform loads. Stirrup

placement will be briefly examined in the design of the point-loaded girders in the following section.

16.4 GIRDER DESIGN

Figure 16.9 shows one of the two interior continuous girders running north to south at the third floor level. As mentioned previously, the effects of openings in the floor slab will be ignored in determining the beam and girder loads.

The girder carries a small portion of exterior wall on its cantilever, but most of its load is floor load, coming in from the beam reactions. It also must carry a small strip of uniform live load directly above its own 16 in width (which was not included in the tributary area for the beams) and, of course, its own weight.

Figure 16.10 shows a continuous beam line (from the 33-ft bay of the girder) delivering its reactions as point loads on the girders. It is important to realize that the continuity in the beam increases the moment and therefore the reactions at the interior supports of the exterior beam spans. The normal reaction is half the load, or

$$\frac{w_u L}{2} = \frac{3.4(28.7)}{2} = 48.8 \approx 49 \text{ kips}$$

However, the ACI coefficients from Figure 4-18*a* indicate that a conservative value for the shear force at the interior support of an exterior span will be 15% larger. Thus, we get

$$1.15 \frac{w_u L}{2} = 1.15 \left[\frac{3.4(28.7)}{2} \right] = 56.1 \approx 56 \text{ kips}$$

FIGURE 16.9
Elevation of interior
third floor girder.

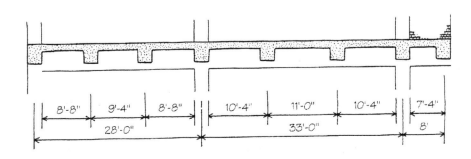

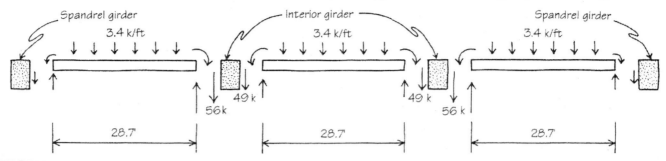

FIGURE 16.10   Continuous beam reactions upon the girders.

The total point load on the girder will be the sum of 55 and 63, or 118 kips as shown in Figure 16.11.

The reactions from the beams in the 28-ft bays will be less because the beams are more closely spaced. Reducing the previous values by using a ratio of tributary strips for the two different spacings, we obtain

$$\frac{9.3}{11.0}(50) + \frac{9.3}{11.0}(56) =$$

$$43 \quad + \quad 47 = 90 \text{ kips}$$

(For the beams that intersect the girder at the column line these reactions will load the column directly and not influence the girder design.)

The spandrel beam reaction at the end of the cantilever will be quite small because of its very small tributary area. The weight of the glass wall on the beam is negligible, so an estimate for the reaction can be based simply upon the tributary strip (4 ft) of floor slab. Again, using a ratio of this tributary strip to that of an interior beam, we get

$$\frac{4.0}{11.0}(49) = 18 \text{ kips}$$

(Note that this load will come into the girder from one side only.)

The overall thickness h of the girder is likely to be less for the 28-ft bay than for the 33-ft bay. For the purpose of including the self-weight let us assume that the thickness is constant for the entire girder. An estimate of h = 38 in will be tried. Its weight can then be found as 633 plf. The previously mentioned strip of live load directly above the girder is 16 in wide (1.33 ft), so the uniform live load will be 1.33 × 100, or 133 plf. The total

FIGURE 16.11   Estimated loads on an interior third floor girder.

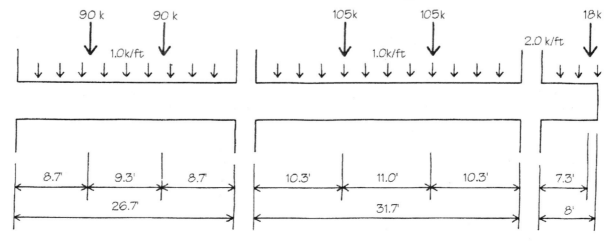

factored uniform load is then

$$w_u = 1.6(133) + 1.2(633)$$

$$= 972 \text{ plf} \approx 1.0 \text{ klf}$$

The exterior masonry wall on the cantilever weighs 80 lb per square foot of elevation (Table 16.1) and there is approximately 10 ft of wall vertically to the underside of the roof girder, so the factored wall load will be

$$1.2(80 \times 10) = 960 \text{ plf} \approx 1.0 \text{ klf}$$

The total loading pattern appears as in Figure 16.11. (We recognize that the uniform load on the cantilever is conservative in that it includes both the masonry wall and the live load, which essentially occupy the same floor space; $w_u = 1.0 + 1.0 = 2.0$ klf.)

As with the beam design of the previous section, an estimate of the maximum moment must be made. The cantilever is, of course, fully determinate and its moment will be

$$M_u = \frac{w_u L^2}{2} + P_u L$$

$$= \frac{2.0(8)^2}{2} + 18(7.3)$$

$$= 195 \text{ kip-ft}$$

Much larger moments are likely to be found in the other two spans. Referring to Figure 4.18b again, the ACI coefficients show that the negative moments at the interior column will control. For the uniform load a value of $wL^2/9$ is conservative. For the point loads Table D of the Appendix provides a negative moment value of $2PL/9$ for a fixed-end beam with third-point loads. For our case the controlling moments at the interior column should be *larger* because the outer supports are less than fixed. On the other hand, the point loads are located outside the middle third points of the girder spans, which should *reduce* all moment values somewhat. Using this reasoning, perhaps $2PL/9$ is not a bad guess for maximum point load negative moments. The controlling moment occurs at the left end of the larger span and is

$$M_u = \frac{w_u L^2}{9} + \frac{2P_u L}{9}$$

$$= \frac{1.0(31.7)^2}{9} + \frac{2(105)(31.7)}{9}$$

$$= 851 \text{ kip-ft} \approx 850 \text{ kip ft}$$

(This value and those that follow are rounded to avoid a pretense of unwarranted accuracy.)

The moment at the right end of the 26.7-ft span will be

$$M_u = \frac{w_u L^2}{9} + \frac{2P_u L}{9}$$

$$= \frac{1.0(26.7)^2}{9} + \frac{2(90)(26.7)}{9}$$

$$= 610 \text{ kip-ft}$$

These and other values appear in Figure 16.12.

FIGURE 16.12   Girder negative moments.

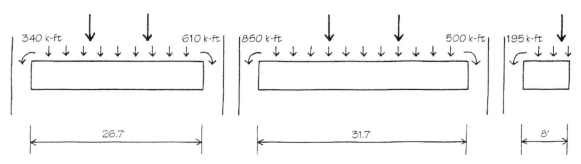

The smaller moments at the outer ends of the girder spans were obtained by using a ratio of the ACI moment coefficients, $wL^2/16$ and $wL^2/9$, from Figure 4.18b:

$$\frac{1/16}{1/9} = 0.56$$

The moment at the left-hand column is then

$$0.56(610) \approx 340 \text{ kip-ft}$$

and at the right-hand column, we get

$$0.56(850) = 476 \text{ kip-ft}$$

This value will be increased (somewhat arbitrarily) to 500 kip-ft to account for the added resistance to joint rotation provided by the cantilever.

Presuming the same design p value that we used for the beam ($\rho = \frac{2}{3}\rho_{max} = 0.0142$), the R value from Table B.1(4) will again be 0.745. The required effective depth then is

$$d = \sqrt{\frac{M_u}{\phi b R}} \qquad (7\text{-}2b)$$

$$= \sqrt{\frac{850(12)}{0.9(16)(0.745)}}$$

$$= 30.8 \approx 31 \text{ in}$$

The amount of steel needed will be

$$A_s = \rho b d \qquad (6\text{-}7)$$

$$= 0.0142(16)(31)$$

$$= 7.04 \text{ in}^2$$

Table A.1 shows that six #10 bars ($A_s = 7.59$ in^2) *or* eight #9 bars ($A_s = 8.00$ in^2) will do the job. Table A.3 shows that in either case two layers have to be used (or since this is top steel, the outer bars of a single layer can be permitted to lie in the slab just beyond the beam's stem width). Figure 16.13 shows two layers of #9 bars. Following the spacing requirements given in Figure 2.9 and assuming ¾-in maximum aggregate size, we can see that our estimate of 38 in for the overall depth was sufficient. If the detailing is done prop-

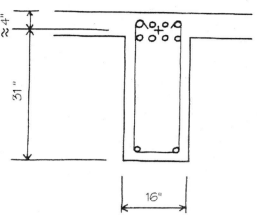

FIGURE 16.13   Girder steel at the interior column.

erly, the single layer of steel at the top of the beam will fit between the two layers of girder steel when the two members intersect at a column.

The overall girder dimensions will not be greater than 16 × 35 and, as mentioned previously, the required depth of the 26.7-ft span (and that of the cantilever) will be less.

16.5 SHEAR IN THE GIRDER

Even though it is unlikely that shear will control the overall size of a bending member, the presence of point loads makes the girder a candidate for a brief check on the required spacing of stirrups. The maximum shear values will occur at the girder ends as reactions and are shown in Figure 16.14. These may be obtained from statics and the largest one occurs at the left end of the 31.7-ft span girder and is

$$R_L = \frac{1}{2}(1.0)(31.7) + 105 + \frac{850 - 500}{31.7}$$

$$= 131.9 \approx 132 \text{ kips}$$

(A more conservative approach to obtaining the large interior reactions would involve applying the ACI 15% increase to one-half the load on each span as we did with the beam in obtaining the girder loads. Due to the stiffness

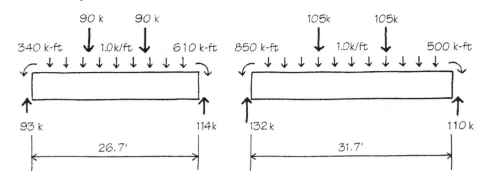

FIGURE 16.14
Approximate girder maximum shear value (reactions).

provided to the outer ends of the girder by the columns, we believe that this degree of con servatism is not warranted for the girder.)

Only a small amount of shear reduction is possible since the slope of the V diagram is so small. Taking it anyway, Figure 16.15 shows the critical shear force to be

$$V_u = 132 - 2.58(1.0) = 129 \text{ kips}$$

Using the procedures of Chapter 8, we can establish the required spacing of stirrups for the critical shear value:

$$V_c = 2\sqrt{f_c'}bd \qquad (8\text{-}2)$$

$$= 2\sqrt{4000}(16)(31)$$

$$= 62\,740 \text{ lb} \approx 62.8 \text{ kips}$$

FIGURE 6.15

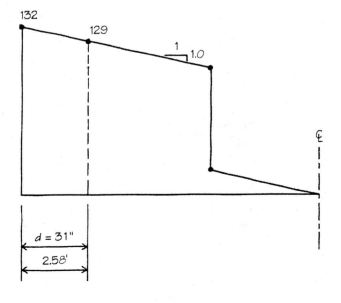

$$V_s = \frac{V_u}{\phi} - V_c \qquad (8\text{-}1a)$$

$$= \frac{129}{0.75} - 62.8$$

$$= 109 \text{ kips}$$

Using Grade 60 steel and #4 stirrups with an area of 0.40 in^2 (0.20 in^2 for each leg), we get

$$s = \frac{A_v f_y d}{V_s}$$

$$= \frac{0.40(60)(31)}{109}$$

$$= 6.8 \text{ in}$$

This is a reasonable stirrup spacing, and it is clear that the girder size will not have to be increased because of shear.

16.6 COLUMN SIZE

Our beginning estimate of column size was 16 × 16 in. In terms of axial loads, the most heavily loaded columns are the interior ones such as those at B-2 and C-2 in Figure 16.2. Either of these must support two girder reactions and two beam reactions at each level. (Those reaction values will be smaller at the roof level, of course, because of the much smaller loads involved.)

For each of the two above-grade floors the load from the girders (to the north and south of column B-2, e.g.) will be the reaction values from Figure 16.14, i.e., 132 and 114 kips. The

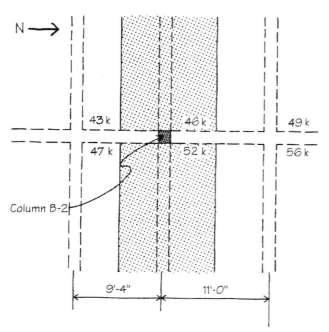

FIGURE 16.16 Beam reaction values.

beams that frame into this column from the east and west lie between the two girder bays, and thus will have tributary slab areas that lie partly in each bay. Figure 16.16 shows this and the various beam reaction values. The reactions shown for the beams that frame into the girder are those obtained in Section 16.4. The reactions for the beams framing directly into the column were taken as halfway be tween these values, reflecting the shared tributary area. (The alert reader will notice that the 52-kip value in Figure 16.14 is probably conservative, because it is the rounded average of the 47- and 56-kip values that exist at the interior ends of beams whose opposite ends frame into a spandrel girder. These values include a 15% increase, which should probably be reduced for a beam whose opposite end frames into a column.)

As previously mentioned, the loads from the roof will be smaller. Since the applied loads are only about one-third those acting on the other levels, the member sizes will all be smaller. To avoid having to design the roof members just to get column loads, let us take the beam and girder reactions at the roof level

as a reasonable percentage, say, 60%, of those at one of the floors. At first this may seem to be overly conservative in view of the relatively small applied loads at the roof level. However, a few simple calculations will show that about one-half the total floor load, which we just tabulated, came from structural self-weight. Furthermore, we know that the thickness of structural elements does not vary linearly with load and moment. In the basic moment equation for strength design, e.g., d varies as the square root of the moment. In other words, a 50% reduction in moment (which does vary linearly with load) may only result in a 25% reduction in depth.

Using the 60% values at the roof, we get the column loads shown in Figure 16.17. These total to 894 kips, acting on the first

FIGURE 16.17 Estimated loads on column B-2.

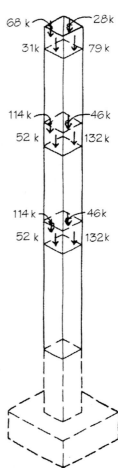

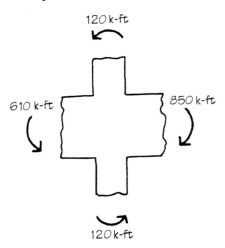

FIGURE 16.18   Joint equilibrium.

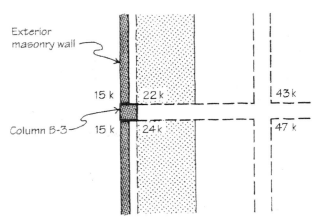

FIGURE 16.19   Beam reaction values.

story segment. There will be very little moment applied to this column by the beams. The girders, however, have different end moments of 850 and 610 kip-ft, as illustrated in Figure 16.12. (These values are for a third floor girder, but the end moments for a second floor girder will be about the same.) The column segments above and below the girders have to absorb the difference between these girder moments in order to achieve joint equilibrium. Figure 16.18 shows how column B-2 must carry 120 kip-ft of moment at each floor level.

Our design loads are, therefore, $P_u = 894$ kips and $M_u = 120$ kip-ft. (A quick check indicates that the column self-weight is negligible.) Graph C.2 (6%) shows us that a 16 × 16-in section with 6% steel placed symmetrically on all four sides will be adequate. (The use of live load reduction and refinements in our conservative design loads during the final design would probably show that less steel would suffice.)

An exterior column such as B-3 will have much less axial load but more moment. From Figure 16.14, the reaction at the left end of the girder is 93 kips. The interior beams in the 28-ft bay have reactions of 43 kips from the west and 47 kips from the east as seen in Figure 16.14. However, the spandrel beams which frame into column B-3 will have only half that

much load because they have only half as much tributary area (Figure 16.19). They do, however, carry the 80-psf exterior wall for their full 28 ft-8 in-span lengths (assuming there are no windows). If the distance is 11 ft to the underside of the beam above, the factored load of the wall on the column from the east or the west will be

$$1.2\left(80 \times 11 \times \frac{28.7}{2}\right) = 15\,200\text{ lb} \approx 15\text{ kips}$$

These values are also shown in Figure 16.19. As before, if we estimate the roof deck loads at 60% of the floor loads, we will get the column loads in Figure 16.20. (There is no wall load at the roof level, of course, so the beam reactions there represent 60% of the floor loads only.)

As we saw with column B-2, the columns above and below the girder will have to share the girder end moment, which (from Figure 16.12) has a value of 340 kip-ft. Figure 16.21 shows how each column will receive 170 kip-ft from the girder.

Summing up the loads on B-3, we get design loads of $P_u = 421$ kips and $M_u = 170$ kip-ft. Graph C.2 (4%) shows that a 16-in-square column with 4% of steel will be more than adequate for these loads.

At this point it is advisable to remember, as stated in Section 16.1, that this analysis does not include the effects of lateral forces. In

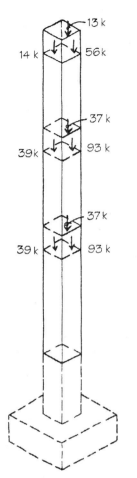

FIGURE 16.20 Estimated loads on column B-3.

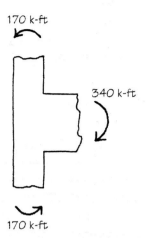

FIGURE 16.21 Joint equilibrium.

accommodating such forces, there is little doubt that the amount and placement of steel in the beams, girders, and columns will be influenced.

16.7 FOOTING DESIGN

The largest column pads occur under the interior columns, B-2 and C-2. Their plan dimensions are determined by the magnitude of the *service* loads and the effective allowable pressure of the soil. Following the procedures of Chapter 13, footing thickness is determined by punching shear considerations using strength design procedures.

The total service load acting on B-2 or C-2 cannot be determined merely by adding up the applied loads and structural self-weight on the slabs, beams, and girders within the column's tributary area at each level. This would suffice if the building were larger in plan and these columns were located more than one bay away from the building exterior. However, as seen in the previous sections of this chapter, these particular columns receive a greater portion of the load due to the larger negative moment at the first interior support of any continuous beam or girder. Since the total factored load on column B-2 (894 kips from Figure 16.17) includes this effect, we should be able to get a reasonable estimate of the service load by "unfactoring," i.e., dividing the dead loads by 1.2 and the live loads by 1.6. This process is greatly simplified if we take advantage of the previously observed fact that about half our total load comes from the structure itself. Since only a small portion of the applied loads is dead (Table 16.1), we can safely say that somewhat more than half the 894 kips comes from the dead load. Since a load factor of 1.4 (halfway between 1.2 and 1.6) will be valid if the load were 50% live and 50% dead, then using a factor of 1.35 will be appropriate for our situation. Thus, a reasonable estimate of the service load on column B-2 will be

$$\frac{894}{1.35} = 662 \text{ kips}$$

To get the effective allowable soil bearing pressure q_e we need to reduce the soil bearing capacity of 5500 psf (given in Table 16.1) by all the loads that act on the soil other than through the column. This will include the first floor applied load, the first floor slab-on-grade, the soil between the slab and the footing, and the footing itself. To do this, let us assume a slab-on-grade thickness of 5 in and footing thickness of 32 in. (This implies an effective depth of about 27 in, allowing for 3 in of clear cover and space for two layers of steel.) We can then find q_e by referring to Figure 16.22.

Soil bearing capacity:	
(Table 16.1)	5500 psf
Applied load:	
100 live, 17 dead	
(Table 16.1)	117
Slab-on-grade:	
5 in thick	63
Soil: 3 ft thick	
(Figure 16.3)	
@ 100 pcf	
(Table 16.1)	300
Footing:	
2 ft-8 in thick	400
$q_e =$	4620 psf
	= 4.62 ksf

The required footing area will be

$$A_r = \frac{P}{q_e}$$

$$= \frac{662}{4.62}$$

$$= 143.3 \text{ ft}^2$$

A 12 × 12-ft-square pad will be adequate. (This may seem large for this size of building but remember that we did not take a live load reduction as permitted by building codes.)

Next the assumed depth of the footing will be checked to see if it is safe against punching shear. The upward pressure on the bottom of

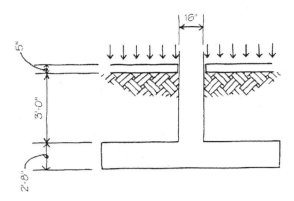

FIGURE 16.22

the footing due to the factored load is

$$q_e = \frac{894}{144}$$

$$= 6.2 \text{ ksf} = 6200 \text{ psf}$$

Referring to Figure 13.5 but using our dimensions, we can find that the plan area of the footing, which develops the punching shear force, is as shown in Figure 16.23.

The ultimate punching shear force will be

$$V_u = 6200\left[144 - \left(\frac{16+27}{12}\right)^2\right]$$

$$= 813\,000 \text{ lb}$$

FIGURE 16.23   Punching shear plan area.

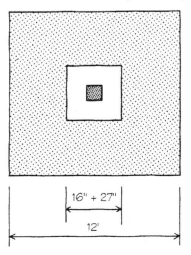

From Table 16.1, the strength of the concrete is 3500 psi. We can use Equation 13-1 to find the capacity of this footing in punching shear:

$$V_c = \phi 4\sqrt{f_c'} b_o d$$

The perimeter of the punching shear surface is

$$b_o = 4(16 + 27) = 172 \text{ in}$$

Therefore,

$$V_c = 0.75(4)\sqrt{3500}(172)(27)$$

$$= 824\,000 \text{ lb}$$

Since V_c is greater than V_u, the footing thickness is adequate.

Now we need to determine the amount of steel needed for moment. The length of the inverted cantilever is $\left(12 - \frac{16}{12}\right)/2$ or 5.33 ft, as shown in Figure 16.24. Its width is the footing width of 12 ft.

The maximum moment is

$$M_u = \frac{w_u L^2}{2}$$

$$= \frac{74.4(5.33)^2}{2}$$

$$= 1057 \text{ kip-ft}$$

FIGURE 16.24

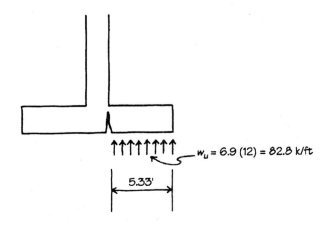
$w_u = 6.9\,(12) = 82.8 \text{ k/ft}$
5.33'

The needed coefficient of resistance will be

$$\frac{M_r}{\phi b d^2} = R$$

$$\frac{1057(12)}{0.9(144)(27)^2} = 0.134$$

Using Table B.4, we note that a concrete strength between 3000 and 4000 psi will yield a ρ value of 0.0023.

$$A_s = \rho b d$$

$$= 0.0023(144)(27)$$

$$= 7.77 \text{ in}^2$$

Allowing for the required cover, the bars must fit into a footing width of about 136 inches. This means at least nine bars (eight spaces) to meet the maximum spacing permitted of 18 inches. Nine #9 bars in each direction will be more than adequate.

16.8 CONCLUSIONS

As stated at the beginning of this chapter, the purpose of doing this preliminary design work was to check the feasibility of the structural scheme. The member sizes we obtained were in some cases quite large. This was due, in part, to the fact that conservative assumptions were made with respect to the effects of continuity.

The girders, columns, and footings could also have benefitted from the proper use of a live load reduction provision permitted by most codes, as mentioned at the beginning of this chapter. The column and its footing that were designed in the last section have especially large tributary areas and thus would qualify for large reductions in the live loads.

Several inches could be pared from the depths of both beams and girders simply by using a larger steel ratio. The girders, at their current overall thickness of 35 in, constitute a particular problem because there would be insufficient clearance for any east to west duct runs that must pass beneath them. (One of the reasons we prefer using small steel ratios for preliminary design purposes is so that such problems could be easily addressed at a later time without having to make changes in the overall framing scheme.)

Viewing the excess member sizes from a more holistic standpoint, one could examine the architectural schematics to check the feasibility of reducing the girder spans. Without changing the beam locations or stairwell placement, would it be possible, for instance, to use four columns along each girder line instead of three? This would reduce the maximum girder span to 22 ft and could result in girder depths as small as the beam depths in each bay. It would then be likely that the floor-to-floor height could be reduced from 13 to 12 ft, or even less, and still leave enough mechanical space.

The plan would have to accommodate four interior columns instead of two, but the architectural character of the building would be changed very little and significant savings (both initial and long term) could result from the increase in structural efficiency.

APPENDIX

TABLE A.1 *Areas of Groups of Bars (in²)*

Bar Number		Number of Bars									
In-lb	Metric*	1	2	3	4	5	6	7	8	9	10
3†	#10	0.11	0.22	0.33	0.44	0.55	0.66	0.77	0.88	0.99	1.10
4†	#13	0.20	0.39	0.58	0.78	0.98	1.18	1.37	1.57	1.77	1.96
5	#16	0.31	0.61	0.91	1.23	1.54	1.85	2.16	2.47	2.78	3.09
6	#19	0.44	0.88	1.32	1.76	2.20	2.64	3.09	3.52	3.97	4.41
7	#22	0.60	1.20	1.80	2.40	3.01	3.60	4.20	4.80	5.40	6.00
8	#25	0.79	1.57	2.35	3.16	3.93	4.72	5.51	6.30	7.09	7.88
9	#29	1.00	2.00	3.00	4.00	5.00	6.00	7.00	8.00	9.00	10.00
10	#32	1.27	2.53	3.79	5.06	6.33	7.59	8.86	10.12	11.39	12.66
11	#36	1.56	3.12	4.68	6.24	7.80	9.36	10.92	12.48	14.04	15.60
12‡	#43	2.25	4.50	6.75	9.00	11.25	13.50	15.75	18.00	20.25	22.50
13‡	#57	4.00	8.00	12.00	16.00	20.00	24.00	28.00	32.00	36.00	40.00

*At this writing, with the exception of work for the federal government, most practitioners use the inch-pound system. The metric system sizes are provided here because of the CRSI (See Section 7.6) mandates that reinforcing bars carry only metric markings.

† These sizes are generally used in slabs or as stirrups or ties.
‡ These sizes are generally used only in columns.

TABLE A.2 ρ_b, ρ_{max}, and ρ_{min} for Selected Combinations of Concrete and Steel

f_y = 40 ksi				f_y = 60 ksi			
f'_c (psi)	ρ_b	ρ_{max}	ρ_{min}	f'_c (psi)	ρ_b	ρ_{max}	ρ_{min}
3000	0.0371	0.0278	0.0050	3000	0.0214	0.0160	0.0033
4000	0.0495	0.0371	0.0050	4000	0.0285	0.0213	0.0033
5000	0.0582	0.0436	0.0050	5000	0.0335	0.0251	0.0033

TABLE A.3 *Minimum Beam Widths for Bars in One Layer* † ‡ *(in)*

Bar No.	Number of Bars						
	2	**3**	**4**	**5**	**6**	**7**	**8**
4	7.0	8.5	10.0	11.5	13.0	14.5	16.0
5	7.1	8.7	10.4	12.0	13.6	15.2	16.9
6	7.2	9.0	10.7	12.5	14.2	16.0	17.7
7	7.4	9.2	11.1	13.0	14.9	16.7	18.6
8	7.5	9.5	11.5	13.5	15.5	17.5	19.5
9	7.7	10.0	12.4	14.5	16.8	19.0	21.3
10	8.0	10.6	13.1	15.7	18.2	20.7	23.3
11	8.3	11.1	14.0	16.8	19.6	22.4	25.2

† Assumes #4 stirrups, ¾-in maximum aggregate size, and 1½ in of clear cover.
‡ Overall beam widths are usually detailed to the next larger whole inch.

TABLES B.1   ***Coefficient of Resistance R (ksi)***

Table B.1   (3)

$f_c' = 3000$ psi $f_y = 60\,000$ psi

ρ	R	ρ	R	ρ	R	ρ	R	ρ	R
0.0033	0.190	0.0059	0.329	0.0085	0.459	0.0111	0.579	0.0137	0.689
0.0034	0.196	0.0060	0.335	0.0086	0.464	0.0112	0.583	0.0138	0.693
0.0035	0.201	0.0061	0.340	0.0087	0.468	0.0113	0.588	0.0139	0.697
0.0036	0.207	0.0062	0.345	0.0088	0.473	0.0114	0.592	0.0140	0.701
0.0037	0.212	0.0063	0.350	0.0089	0.478	0.0115	0.596	0.0141	0.705
0.0038	0.218	0.0064	0.355	0.0090	0.483	0.0116	0.601	0.0142	0.709
0.0039	0.223	0.0065	0.360	0.0091	0.487	0.0117	0.605	0.0143	0.713
0.0040	0.229	0.0066	0.365	0.0092	0.492	0.0118	0.609	0.0144	0.717
0.0041	0.234	0.0067	0.370	0.0093	0.497	0.0119	0.614	0.0145	0.721
0.0042	0.240	0.0068	0.375	0.0094	0.501	0.0120	0.618	0.0146	0.725
0.0043	0.245	0.0069	0.380	0.0095	0.506	0.0121	0.622	0.0147	0.729
0.0044	0.250	0.0070	0.385	0.0096	0.511	0.0122	0.627	0.0148	0.733
0.0045	0.256	0.0071	0.390	0.0097	0.515	0.0123	0.631	0.0149	0.737
0.0046	0.261	0.0072	0.395	0.0098	0.520	0.0124	0.635	0.0150	0.741
0.0047	0.266	0.0073	0.400	0.0099	0.525	0.0125	0.639	0.0151	0.745
0.0048	0.272	0.0074	0.405	0.0100	0.529	0.0126	0.644	0.0152	0.748
0.0049	0.277	0.0075	0.410	0.0101	0.534	0.0127	0.648	0.0153	0.752
0.0050	0.282	0.0076	0.415	0.0102	0.538	0.0128	0.652	0.0154	0.756
0.0051	0.288	0.0077	0.420	0.0103	0.543	0.0129	0.656	0.0155	0.760
0.0052	0.293	0.0078	0.425	0.0104	0.547	0.0130	0.660	0.0156	0.764
0.0053	0.298	0.0079	0.430	0.0105	0.552	0.0131	0.665	0.0157	0.767
0.0054	0.303	0.0080	0.435	0.0106	0.556	0.0132	0.669	0.0158	0.771
0.0055	0.309	0.0081	0.440	0.0107	0.561	0.0133	0.673	0.0159	0.775
0.0056	0.314	0.0082	0.444	0.0108	0.565	0.0134	0.677	0.0160	0.779
0.0057	0.319	0.0083	0.449	0.0109	0.570	0.0135	0.681		
0.0058	0.324	0.0084	0.454	0.0110	0.574	0.0136	0.685		

TABLES B.1   *Coefficient of Resistance R (ksi)*

Table B.1   (4)

$f'_c = 4000$ psi $f_y = 60\ 000$ psi

ρ	R	ρ	R	ρ	R	ρ	R	ρ	R
0.0033	0.192	0.0070	0.394	0.0107	0.581	0.0144	0.754	0.0181	0.912
0.0034	0.198	0.0071	0.399	0.0108	0.586	0.0145	0.758	0.0182	0.916
0.0035	0.203	0.0072	0.404	0.0109	0.591	0.0146	0.763	0.0183	0.920
0.0036	0.209	0.0073	0.410	0.0110	0.596	0.0147	0.767	0.0184	0.924
0.0037	0.215	0.0074	0.415	0.0111	0.601	0.0148	0.772	0.0185	0.928
0.0038	0.220	0.0075	0.420	0.0112	0.605	0.0149	0.776	0.0186	0.932
0.0039	0.226	0.0076	0.425	0.0113	0.610	0.0150	0.781	0.0187	0.936
0.0040	0.232	0.0077	0.431	0.0114	0.615	0.0151	0.785	0.0188	0.940
0.0041	0.237	0.0078	0.436	0.0115	0.620	0.0152	0.789	0.0189	0.944
0.0042	0.243	0.0079	0.441	0.0116	0.625	0.0153	0.794	0.0190	0.948
0.0043	0.248	0.0080	0.446	0.0117	0.629	0.0154	0.798	0.0191	0.952
0.0044	0.254	0.0081	0.451	0.0118	0.634	0.0155	0.802	0.0192	0.956
0.0045	0.259	0.0082	0.456	0.0119	0.639	0.0156	0.807	0.0193	0.960
0.0046	0.265	0.0083	0.461	0.0120	0.644	0.0157	0.811	0.0194	0.964
0.0047	0.270	0.0084	0.467	0.0121	0.648	0.0158	0.815	0.0195	0.968
0.0048	0.276	0.0085	0.472	0.0122	0.653	0.0159	0.820	0.0196	0.972
0.0049	0.281	0.0086	0.477	0.0123	0.658	0.0160	0.824	0.0197	0.976
0.0050	0.287	0.0087	0.482	0.0124	0.662	0.0161	0.828	0.0198	0.980
0.0051	0.292	0.0088	0.487	0.0125	0.667	0.0162	0.833	0.0199	0.984
0.0052	0.298	0.0089	0.492	0.0126	0.672	0.0163	0.837	0.0200	0.988
0.0053	0.303	0.0090	0.497	0.0127	0.676	0.0164	0.841	0.0201	0.991
0.0054	0.309	0.0091	0.502	0.0128	0.681	0.0165	0.845	0.0202	0.995
0.0055	0.314	0.0092	0.507	0.0129	0.686	0.0166	0.850	0.0203	0.999
0.0056	0.319	0.0093	0.512	0.0130	0.690	0.0167	0.854	0.0204	1.003
0.0057	0.325	0.0094	0.517	0.0131	0.695	0.0168	0.858	0.0205	1.007
0.0058	0.330	0.0095	0.522	0.0132	0.699	0.0169	0.862	0.0206	1.011
0.0059	0.336	0.0096	0.527	0.0133	0.704	0.0170	0.867	0.0207	1.014
0.0060	0.341	0.0097	0.532	0.0134	0.709	0.0171	0.871	0.0208	1.018
0.0061	0.346	0.0098	0.537	0.0135	0.713	0.0172	0.875	0.0209	1.022
0.0062	0.352	0.0099	0.542	0.0136	0.718	0.0173	0.879	0.0210	1.026
0.0063	0.357	0.0100	0.547	0.0137	0.722	0.0174	0.883	0.0211	1.030
0.0064	0.362	0.0101	0.552	0.0138	0.727	0.0175	0.887	0.0212	1.033
0.0065	0.368	0.0102	0.557	0.0139	0.731	0.0176	0.892	0.0213	1.037
0.0066	0.373	0.0103	0.562	0.0140	0.736	0.0177	0.896		
0.0067	0.378	0.0104	0.567	0.0141	0.740	0.0178	0.900		
0.0068	0.383	0.0105	0.571	0.0142	0.745	0.0179	0.904		
0.0069	0.389	0.0106	0.576	0.0143	0.749	0.0180	0.908		

TABLES B.1   *Coefficient of Resistance R (ksi)*

Table B.1   (5)

f'_c = 5000 psi f_y = 60 000 psi

ρ	R	ρ	R	ρ	R	ρ	R	ρ	R
0.0033	0.193	0.0077	0.437	0.0121	0.664	0.0165	0.874	0.0209	1.068
0.0034	0.199	0.0078	0.442	0.0122	0.669	0.0166	0.879	0.0210	1.073
0.0035	0.205	0.0079	0.447	0.0123	0.674	0.0167	0.884	0.0211	1.077
0.0036	0.210	0.0080	0.453	0.0124	0.679	0.0168	0.888	0.0212	1.081
0.0037	0.216	0.0081	0.458	0.0125	0.684	0.0169	0.893	0.0213	1.085
0.0038	0.222	0.0082	0.463	0.0126	0.689	0.0170	0.897	0.0214	1.089
0.0039	0.228	0.0083	0.469	0.0127	0.693	0.0171	0.902	0.0215	1.094
0.0040	0.233	0.0084	0.474	0.0128	0.698	0.0172	0.906	0.0216	1.098
0.0041	0.239	0.0085	0.479	0.0129	0.703	0.0173	0.911	0.0217	1.102
0.0042	0.245	0.0086	0.485	0.0130	0.708	0.0174	0.915	0.0218	1.106
0.0043	0.250	0.0087	0.490	0.0131	0.713	0.0175	0.920	0.0219	1.110
0.0044	0.256	0.0088	0.495	0.0132	0.718	0.0176	0.924	0.0220	1.114
0.0045	0.261	0.0089	0.500	0.0133	0.723	0.0177	0.929	0.0221	1.119
0.0046	0.267	0.0090	0.506	0.0134	0.728	0.0178	0.933	0.0222	1.122
0.0047	0.273	0.0091	0.511	0.0135	0.733	0.0179	0.938	0.0223	1.127
0.0048	0.278	0.0092	0.516	0.0136	0.737	0.0180	0.942	0.0224	1.131
0.0049	0.284	0.0093	0.521	0.0137	0.742	0.0181	0.947	0.0225	1.135
0.0050	0.289	0.0094	0.526	0.0138	0.747	0.0182	0.951	0.0226	1.139
0.0051	0.295	0.0095	0.532	0.0139	0.752	0.0183	0.956	0.0227	1.143
0.0052	0.301	0.0096	0.537	0.0140	0.757	0.0184	0.960	0.0228	1.147
0.0053	0.306	0.0097	0.542	0.0141	0.762	0.0185	0.965	0.0229	1.151
0.0054	0.312	0.0098	0.547	0.0142	0.766	0.0186	0.969	0.0230	1.155
0.0055	0.317	0.0099	0.552	0.0143	0.771	0.0187	0.973	0.0231	1.159
0.0056	0.323	0.0100	0.558	0.0144	0.776	0.0188	0.978	0.0232	1.168
0.0057	0.328	0.0101	0.563	0.0145	0.781	0.0189	0.982	0.0233	1.167
0.0058	0.334	0.0102	0.568	0.0146	0.785	0.0190	0.987	0.0234	1.171
0.0059	0.339	0.0103	0.573	0.0147	0.790	0.0191	0.991	0.0235	1.175
0.0060	0.345	0.0104	0.578	0.0148	0.795	0.0192	0.995	0.0236	1.179
0.0061	0.350	0.0105	0.583	0.0149	0.800	0.0193	1.000	0.0237	1.183
0.0062	0.356	0.0106	0.588	0.0150	0.804	0.0194	1.004	0.0238	1.187
0.0063	0.361	0.0107	0.593	0.0151	0.809	0.0195	1.008	0.0239	1.191
0.0064	0.367	0.0108	0.598	0.0152	0.814	0.0196	1.013	0.0240	1.195
0.0065	0.372	0.0109	0.604	0.0153	0.819	0.0197	1.017	0.0241	1.199
0.0066	0.377	0.0110	0.609	0.0154	0.823	0.0198	1.021	0.0242	1.203
0.0067	0.383	0.0111	0.614	0.0155	0.828	0.0199	1.026	0.0243	1.207
0.0068	0.388	0.0112	0.619	0.0156	0.833	0.0200	1.030	0.0244	1.211
0.0069	0.394	0.0113	0.624	0.0157	0.837	0.0201	1.034	0.0245	1.215
0.0070	0.399	0.0114	0.629	0.0158	0.842	0.0202	1.039	0.0246	1.219
0.0071	0.405	0.0115	0.634	0.0159	0.847	0.0203	1.043	0.0247	1.223
0.0072	0.410	0.0116	0.639	0.0160	0.851	0.0204	1.047	0.0248	1.227
0.0073	0.415	0.0117	0.644	0.0161	0.856	0.0205	1.051	0.0249	1.231
0.0074	0.421	0.0118	0.649	0.0162	0.861	0.0206	1.056	0.0250	1.234
0.0075	0.426	0.0119	0.654	0.0163	0.865	0.0207	1.060	0.0251	1.238
0.0076	0.431	0.0120	0.659	0.0164	0.870	0.0208	1.064		

TABLES B.2 *M_r (kip-ft) for Sections 12 in wide*

Table B.2   (40/3)

$f_y = 40\ 000$ psi $f_c' = 3000$ psi

ρ	3.0	3.5	4.0	4.5	5.0	5.5	6.0	6.5	7.0	8.0	9.0	10.0	11.0	12.0
					Effective Depth (in)									
0.0020	0.6	0.9	1.1	1.4	1.8	2.1	2.6	3.0	3.5	4.5	5.7	7.1	8.6	10.2
0.0025	0.8	1.1	1.4	1.8	2.2	2.7	3.2	3.7	4.3	5.6	7.1	8.8	10.7	12.7
0.0030	0.9	1.3	1.7	2.1	2.6	3.2	3.8	4.5	5.2	6.7	8.5	10.5	12.8	15.2
0.0035	1.1	1.5	2.0	2.5	3.1	3.7	4.4	5.2	6.0	7.8	9.9	12.3	14.8	17.6
0.0040	1.3	1.7	2.2	2.8	3.5	4.2	5.0	5.9	6.8	8.9	11.3	13.9	16.9	20.1
0.0045	1.4	1.9	2.5	3.2	3.9	4.7	5.6	6.6	7.7	10.0	12.7	15.6	18.9	22.5
0.0050	1.6	2.1	2.8	3.5	4.3	5.2	6.2	7.3	8.5	11.1	14.0	17.3	20.9	24.9
0.0055	1.7	2.3	3.0	3.8	4.7	5.7	6.8	8.0	9.3	12.1	15.3	18.9	22.9	27.3
0.0060	1.9	2.5	3.3	4.2	5.1	6.2	7.4	8.7	10.1	13.2	16.7	20.6	24.9	29.6
0.0065	2.0	2.7	3.6	4-5	5.6	6.7	8.0	9.4	10.9	14.2	18.0	22.2	26.9	32.0
0.0070	2.1	2.9	3.8	4.8	6.0	7.2	8.6	10.1	11.7	15.2	19.3	23.8	28.8	34.3
0.0075	2.3	3.1	4.1	5.1	6.4	7.7	9.1	10.7	12.4	16.3	20.6	25.4	30.7	36.6
0.0080	2.4	3.3	4.3	5.5	6.7	8.2	9.7	11.4	13.2	17.3	21.9	27.0	32.7	38.9
0.0085	2.6	3.5	4.6	5.8	7.1	8.6	10.3	12.1	14.0	18.3	23.1	28.6	34.6	41.1
0.0090	2.7	3.7	4.8	6.1	7.5	9.1	10.8	12.7	14.8	19.3	24.4	30.1	36.4	43.4
0.0095	2.8	3.9	5.1	6.4	7.9	9.6	11.4	13.4	15.5	20.3	25.6	31.6	38.3	45.6
0.0100	3.0	4.1	5.3	6.7	8.3	10.0	11.9	14.0	16.3	21.2	26.9	33.2	40.1	47.8
0.0105	3.1	4.2	5.5	7.0	8.7	10.5	12.5	14.7	17.0	22.2	28.1	34.7	42.0	49.9
0.0110	3.3	4.4	5.8	7.3	9.0	10.9	13.0	15.3	17.7	23.2	29.3	36.2	43.8	52.1
0.0115	3.4	4.6	6.0	7.6	9.4	11.4	13.6	15.9	18.5	24.1	30.5	37.7	45.6	54.2
0.0120	3.5	4.8	6.3	7.9	9.8	11.8	14.1	16.5	19.2	25.0	31.7	39.1	47.3	56.3
0.0125	3.7	5.0	6.5	8.2	10.1	12.3	14.6	17.1	19.9	26.0	32.9	40.6	49.1	58.4
0.0130	3.8	5.1	6.7	8.5	10.5	12.7	15.1	17.8	20.6	26.9	34.0	42.0	50.8	60.5
0.0135	3.9	5.3	7.0	8.8	10.9	13.1	15.6	18.4	21.3	27.8	35.2	43.4	52.6	62.6
0.0140	4.0	5.5	7.2	9.1	11.2	13.6	16.1	18.9	22.0	28.7	36.3	44.8	54.3	64.6
0.0145	4.2	5-7	7.4	9.4	11.6	14.0	16.6	19.5	22.7	29.6	37.5	46.2	56.0	66.6
0.0150	4.3	5.8	7.6	9.6	11.9	14.4	17.1	20.1	23.3	30.5	38.6	47.6	57.6	68.6
0.0155	4.4	6.0	7.8	9.9	12.2	14.8	17.6	20.7	24.0	31.4	39.7	49.0	59.3	70.6
0.0160	4.5	6.2	8.1	10.2	12.6	15.2	18.1	21.3	24.7	32.2	40.8	50.4	60.9	72.5
0.0165	4.7	6.3	8.3	10.5	12.9	15.6	18.6	21.8	25.3	33.1	41.9	51.7	62.5	74.4
0.0170	4.8	6.5	8.5	10.7	13.3	16.0	19.1	22.4	26.0	33.9	42.9	53.0	64.1	76.3
0.0175	4.9	6.7	8.7	11.0	13.6	16.4	19.6	23.0	26.6	34.8	44.0	54.3	65.7	78.2
0.0180	5.0	6.8	8.9	11.3	13.9	16.8	20.0	23.5	27.3	35.6	45.1	55.6	67.3	80.1
0.0185	5.1	7.0	9.1	11.5	14.2	17.2	20.5	24.0	27.9	36.4	46.1	56.9	68.9	81.9
0.0190	5.2	7.1	9.3	11.8	14.5	17.6	20.9	24.6	28.5	37.2	47.1	58.2	70.4	83.8
0.0195	5.3	7.3	9.5	12.0	14.9	18.0	21.4	25.1	29.1	38.0	48.1	59.4	71.9	85.6
0.0200	5.5	7.4	9.7	12.3	15.2	18.4	21.8	25.6	29.7	38.8	49.1	60.7	73.4	87.4
0.0205	5.6	7.6	9.9	12.5	15.5	18.7	22.3	26.2	30.3	39.6	50.1	61.9	74.9	89.1
0.0210	5.7	7.7	10.1	12.8	15.8	19.1	22.7	26.7	30.9	40.4	51.1	63.1	76.4	90.9
0.0215	5.8	7.9	10.3	13.0	16.1	19.5	23.2	27.2	31.5	41.2	52.1	64.3	77.8	92.6
0.0220	5.9	8.0	10.5	13.3	16.4	19.8	23.6	27.7	32.1	41.9	53.0	65.5	79.2	94.3
0.0225	6.0	8.2	10.7	13.5	16.7	20.2	24.0	28.2	32.7	42.7	54.0	66.7	80.7	96.0
0.0230	6.1	8.3	10.9	13.7	17.0	20.5	24.4	28.7	33.2	43.4	54.9	67.8	82.1	97.7
0.0235	6.2	8.4	11.0	14.0	17.2	20.9	24.8	29.1	33.8	44.1	55.9	69.0	83.4	99.3
0.0240	6.3	8.6	11.2	14.2	17.5	21.2	25.2	29.6	34.3	44.9	56.8	70.1	84.8	100.9
0.0245	6.4	8.7	11.4	14.4	17.8	21.5	25.6	30.1	34.9	45.6	57.7	71.2	86.2	102.5
0.0250	6.5	8.9	11.6	14.6	18.1	21.9	26.0	30.5	35.4	46.3	58.6	72.3	87.5	104.1
0.0255	6.6	9.0	11.7	14.9	18.3	22.2	26.4	31.0	36.0	47.0	59.4	73.4	88.8	105.7
0.0260	6.7	9.1	11.9	15.1	18.6	22.5	26.8	31.5	36.5	47.7	60.3	74.5	90.1	107.2
0.0265	6.8	9.3	12.1	15.3	18.9	22.8	27.2	31.9	37.0	48.3	61.2	75.5	91.4	108.7
0.0270	6.9	9.4	12.2	15.5	19.1	23.2	27.6	32.3	37.5	49.0	62.0	76.6	92.6	110.2
0.0275	7.0	9.5	12.4	15.7	19.4	23.5	27.9	32.8	38.0	49.7	62.8	77.6	93.9	111.7

TABLES B.2 M_r *(kip-ft) for Sections 12 in wide*

Table B.2   (40/4)

$$f_y = 40\ 000\ \text{psi} \qquad f_c' = 4000\ \text{psi}$$

ρ	3.0	3.5	4.0	4.5	5.0	5.5	6.0	6.5	7.0	8.0	9.0	10.0	11.0	12.0
					Effective Depth (in)									
0.0020	0.6	0.9	1.1	1.4	1.8	2.2	2.6	3.0	3.5	4.6	5.8	7.1	8.6	10.2
0.0025	0.8	1.1	1.4	1.8	2.2	2.7	3.2	3.7	4.3	5.7	7.2	8.9	10.7	12.8
0.0030	1.0	1.3	1.7	2.1	2.7	3.2	3.8	4.5	5.2	6.8	8.6	10.6	12.8	15.3
0.0035	1.1	1.5	2.0	2.5	3.1	3.7	4.4	5.2	6.0	7.9	10.0	12.3	14.9	17.8
0.0040	1.3	1.7	2.2	2.8	3.5	4.3	5.1	-5.9	6.9	9.0	11.4	14.1	17.0	20.2
0.0045	1.4	1.9	2.5	3.2	3.9	4.8	5.7	6.7	7.7	10.1	12.8	15.8	19.1	22.7
0.0050	1.6	2.1	2.8	3.5	4.4	5.3	6.3	7.4	8.6	11.2	14.1	17.5	21.1	25.2
0.0055	1.7	2.3	3.1	3.9	4.8	5.8	6.9	8.1	9.4	12.3	15.5	19.2	23.2	27.6
0.0060	1.9	2.6	3.3	4.2	5.2	6.3	7.5	8.8	10.2	13.3	16.9	20.8	25.2	30.0
0.0065	2.0	2.8	3.6	4.6	5.6	6.8	8.1	9.5	11.0	14.4	18.2	22.5	27.2	32.4
0.0070	2.2	3.0	3.9	4.9	6.0	7.3	8.7	10.2	11.8	15.5	19.6	24.2	29.2	34.8
0.0075	2.3	3.2	4.1	5.2	6.5	7.8	9.3	10.9	12.6	16.5	20.9	25.8	31.2	37.2
0.0080	2.5	3.4	4.4	5.6	6.9	8.3	9.9	11.6	13.4	17.6	22.2	27.4	33.2	39.5
0.0085	2.6	3.6	4.7	5.9	7.3	8.8	10.5	12.3	14.2	18.6	23.5	29.1	35.2	41.9
0.0090	2.8	3.8	4.9	6.2	7.7	9.3	11.0	13.0	15.0	19.6	24.9	30.7	37.1	44.2
0.0095	2.9	4.0	5.2	6.5	8.1	9.8	11.6	13.6	15.8	20.7	26.1	32.3	39.1	46.5
0.0100	3.0	4.1	5.4	6.9	8.5	10.2	12.2	14.3	16.6	21.7	27.4	33.9	41.0	48.8
0.0105	3.2	4.3	5.7	7.2	8.9	10.7	12.8	15.0	17.4	22.7	28.7	35.5	42.9	51.1
0.0110	3.3	4.5	5.9	7.5	9.3	11.2	13.3	15.6	18.1	23.7	30.0	37.0	44.8	53.3
0.0115	3.5	4.7	6.2	7.8	9.6	11.7	13.9	16.3	18.9	24.7	31.3	38.6	46.7	55.6
0.0120	3.6	4.9	6.4	8.1	10.0	12.1	14.5	17.0	19.7	25.7	32.5	40.1	48.6	57.8
0.0125	3.8	5.1	6.7	8.4	10.4	12.6	15.0	17.6	20.4	26.7	33.8	41.7	50.4	60.0
0.0130	3.9	5.3	6.9	8.8	10.8	13.1	15.6	18.3	21.2	27.7	35.0	43.2	52.3	62.2
0.0135	4.0	5.5	7.2	9.1	11.2	13.5	16.1	18.9	21.9	28.6	36.2	44.7	54.1	64.4
0.0140	4.2	5.7	7.4	9.4	11.6	14.0	16.6	19.5	22.7	29.6	37.5	46.2	55.9	66.6
0.0145	4.3	5.8	7.6	9.7	11.9	14.4	17.2	20.2	23.4	30.5	38.7	47.7	57.8	68.7
0.0150	4.4	6.0	7.9	10.0	12.3	14.9	17.7	20.8	24.1	31.5	39.9	49.2	59.6	70.9
0.0155	4.6	6.2	8.1	10.3	12.7	15.3	18.3	21.4	24.8	32.4	41.1	50.7	61.3	73.0
0.0160	4.7	6.4	8.3	10.6	13.0	15.8	18.8	22.0	25.6	33.4	42.3	52.2	63.1	75.1
0.0165	4.8	6.6	8.6	10.9	13.4	16.2	19.3	22.7	26.3	34.3	43.4	53.6	64.9	77.2
0.0170	5.0	6.7	8.8	11.1	13.8	16.7	19.8	23.3	27.0	35.2	44.6	55.1	66.6	79.3
0.0175	5.1	6.9	9.0	11.4	14.1	17.1	20.3	23.9	27.7	36.2	45.8	56.5	68.4	81.4
0.0180	5.2	7.1	9.3	11.7	14.5	17.5	20.9	24.5	28.4	37.1	46.9	57.9	70.1	83.4
0.0185	5.3	7.3	9.5	12.0	14.8	17.9	21.4	25.1	29.1	38.0	48.1	59.3	71.8	85.4
0.0190	5.5	7.4	9.7	12.3	15.2	18.4	21.9	25.7	29.8	38.9	49.2	60.7	73.5	87.5
0.0195	5.6	7.6	9.9	12.6	15.5	18.8	22.4	26.2	30.4	39.8	50.3	62.1	75.2	89.5
0.0200	5.7	7.8	10.2	12.9	15.9	19.2	22.9	26.8	31.1	40.6	51.4	63.5	76.8	91.4
0.0205	5.8	7.9	10.4	13.1	16.2	19.6	23.4	27.4	31.8	41.5	52.5	64.9	78.5	93.4
0.0210	6.0	8.1	10.6	13.4	16.6	20.0	23.8	28.0	32.5	42.4	53.6	66.2	80.1	95.4
0.0215	6.1	8.3	10.8	13.7	16.9	20.4	24.3	28.6	33.1	43.3	54.7	67.6	81.8	97.3
0.0220	6.2	8.4	11.0	14.0	17.2	20.8	24.8	29.1	33.8	44.1	55.8	68.9	83.4	99.2

TABLES B.2 *(Continued)*

Table B.2 (40/4)

$f_y = 40\ 000$ psi $f'_c = 4000$ psi

ρ	3.0	3.5	4.0	4.5	5.0	5.5	6.0	6.5	7.0	8.0	9.0	10.0	11.0	12.0
					Effective Depth (in)									
0.0225	6.3	8.6	11.2	14.2	17.6	21.2	25.3	29.7	34.4	45.0	56.9	70.2	85.0	101.2
0.0230	6.4	8.8	11.5	14.5	17.9	21.6	25.8	30.2	35.1	45.8	58.0	71.6	86.6	103.1
0.0235	6.6	8.9	11.7	14.8	18.2	22.0	26.2	30.8	35.7	46.6	59.0	72.9	88.2	104.9
0.0240	6.7	9.1	11.9	15.0	18.5	22.4	26.7	31.3	36.3	47.5	60.1	74.2	89.7	106.8
0.0245	6.8	9.2	12.1	15.3	18.9	22.8	27.2	31.9	37.0	48.3	61.1	75.5	91.3	108.6
0.0250	6.9	9.4	12.3	15.5	19.2	23.2	27.6	32.4	37.6	49.1	62.1	76.7	92.8	110.5
0.0255	7.0	9.6	12.5	15.8	19.5	23.6	28.1	33.0	38.2	49.9	63.2	78.0	94.4	112.3
0.0260	7.1	9.7	12.7	16.0	19.8	24.0	28.5	33.5	38.8	50.7	64.2	79.2	95.9	114.1
0.0265	7.2	9.9	12.9	16.3	20.1	24.3	29.0	34.0	39.4	51.5	65.2	80.5	97.4	115.9
0.0270	7.4	10.0	13.1	16.5	20.4	24.7	29.4	34.5	40.0	52.3	66.2	81.7	98.9	117.7
0.0275	7.5	10.2	13.3	16.8	20.7	25.1	29.9	35.0	40.6	53.1	67.2	82.9	100.4	119.4
0.0280	7.6	10.3	13.5	17.0	21.0	25.5	30.3	35.6	41.2	53.9	68.2	84.1	101.8	121.2
0.0285	7.7	10.5	13.7	17.3	21.3	25.8	30.7	36.1	41.8	54.6	69.1	85.3	103.3	122.9
0.0290	7.8	10.6	13.8	17.5	21.6	26.2	31.2	36.6	42.4	55.4	70.1	86.5	104.7	124.6
0.0295	7.9	10.7	14.0	17.8	21.9	26.5	31.6	37.1	43.0	56.1	71.0	87.7	106.1	126.3
0.0300	8.0	10.9	14.2	18.0	22.2	26.9	32.0	37.6	43.6	56.9	72.0	88.9	107.5	128.0
0.0305	8.1	11.0	14.4	18.2	22.5	27.2	32.4	38.0	44.1	57.6	72.9	90.0	109.0	129.7
0.0310	8.2	11.2	14.6	18.5	22.8	27.6	32.8	38.5	44.7	58.4	73.9	91.2	110.3	131.3
0.0315	8.3	11.3	14.8	18.7	23.1	27.9	33.2	39.0	45.2	59.1	74.8	92.3	111.7	132.9
0.0320	8.4	11.4	15.0	18.9	23.4	28.3	33.6	39.5	45.8	59.8	75.7	93.5	113.1	134.6
0.0325	8.5	11.6	15.1	1.9.1	23.6	28.6	34.0	40.0	46.3	60.5	76.6	94.6	114.4	136.2
0.0330	8.6	11.7	15.3	19.4	23.9	28.9	34.4	40.4	46.9	61.2	77.5	95.7	115.8	137.8
0.0335	8.7	11.9	15.5	19.6	24.2	29.3	34.8	40.9	47.4	61.9	78.4	96.8	117.1	139.3
0.0340	8.8	12.0	15.7	19.8	24.5	29.6	35.2	41.3	47.9	62.6	79.3	97.8	118.4	140.9
0.0345	8.9	12.1	15.8	20.0	24.7	29.9	35.6	41.8	48.5	63.3	80.1	98.9	119.7	142.4
0.0350	9.0	12.2	16.0	20.2	25.0	30.2	36.0	42.2	49.0	64.0	81.0	100.0	121.0	144.0
0.0355	9.1	12.4	16.2	20.5	25.3	30.6	36.4	42.7	49.5	64.7	81.8	101.0	122.2	145.5
0.0360	9.2	12.5	16.3	20.7	25.5	30.9	36.7	43.1	50.0	65.3	82.7	102.1	123.5	147.0
0.0365	9.3	12.6	16.5	20.9	25.8	31.2	37.1	43.6	50.5	66.0	83.5	103.1	124.8	148.5
0.0370	9.4	12.8	16.7	21.1	26.0	31.5	37.5	44.0	51.0	66.6	84.3	104.1	126.0	149.9

TABLES B.2 M_r *(kip-ft) for Sections 12 in wide*

Table B.2   (40/5)

$$f_y = 40\ 000\ \text{psi} \qquad f_c' = 5000\ \text{psi}$$

ρ	Effective Depth (in)													
	3.0	3.5	4.0	4.5	5.0	5.5	6.0	6.5	7.0	8.0	9.0	10.0	11.0	12.0
0.0020	0.6	0.9	1.1	1.4	1.8	2.2	2.6	3.0	3.5	4.6	5.8	7.1	8.6	10.3
0.0025	0.8	1.1	1.4	1.8	2.2	2.7	3.2	3.8	4.4	5.7	7.2	8.9	10.8	12.8
0.0030	1.0	1.3	1.7	2.2	2.7	3.2	3.8	4.5	5.2	6.8	8.6	10.6	12.9	15.3
0.0035	1.1	1.5	2.0	2.5	3.1	3.7	4.5	5.2	6.1	7.9	10.0	12.4	15.0	17.8
0.0040	1.3	1.7	2.3	2.9	3.5	4.3	5.1	B.0	6.9	9.0	11.4	14.1	17.1	20.3
0.0045	1.4	1.9	2.5	3.2	4.0	4.8	5.7	6.7	7.8	10.1	12.8	15.9	19.2	22.8
0.0050	1.6	2.2	2.8	3.6	4.4	5.3	6.3	7.4	8.6	11.2	14.2	17.6	21.3	25.3
0.0055	1.7	2.4	3.1	3.9	4.8	5.8	6.9	8.1	9.5	12.3	15.6	19.3	23.3	27.8
0.0060	1.9	2.6	3.4	4.3	5.2	6.3	7.6	8.9	10.3	13.4	17.0	21.0	25.4	30.2
0.0065	2.0	2.8	3.6	4.6	5.7	6.9	8.2	9.6	11.1	14.5	18.4	22.7	27.4	32.7
0.0070	2.2	3.0	3.9	4.9	6.1	7.4	8.8	10.3	11.9	15.6	19.7	24.4	29.5	35.1
0.0075	2.3	3.2	4.2	5.3	6.5	7.9	9.4	11.0	12.8	16.7	21.1	26.0	31.5	37.5
0.0080	2.5	3.4	4.4	5.6	6.9	8.4	10.0	11.7	13.6	17.7	22.4	27.7	33.5	39.9
0.0085	2.6	3.6	4.7	5.9	7.3	8.9	10.6	12.4	14.4	18.8	23.8	29.4	35.5	42.3
0.0090	2.8	3.8	5.0	6.3	7.8	9.4	11.2	13.1	15.2	19.9	25.1	31.0	37.5	44.7
0.0095	2.9	4.0	5.2	6.6	8.2	9.9	11.8	13.8	16.0	20.9	26.5	32.7	39.5	47.0
0.0100	3.1	4.2	5.5	6.9	8.6	10.4	12.3	14.5	16.8	22.0	27.8	34.3	41.5	49.4
0.0105	3.2	4.4	5.7	7.3	9.0	10.9	12.9	15.2	17.6	23.0	29.1	35.9	43.5	51.7
0.0110	3.4	4.6	6.0	7.6	9.4	11.4	13.5	15.9	18.4	24.0	30.4	37.5	45.4	54.1
0.0115	3.5	4.8	6.3	7.9	9.8	11.8	14.1	16.5	19.2	25.1	31.7	39.2	47.4	56.4
0.0120	3.7	5.0	6.5	8.3	10.2	12.3	14.7	17.2	20.0	26.1	33.0	40.8	49.3	58.7
0.0125	3.8	5.2	6.8	8.6	10.6	12.8	15.2	17.9	20.7	27.1	34.3	42.3	51.2	61.0
0.0130	4.0	5.4	7.0	8.9	11.0	13.3	15.8	18.6	21.5	28.1	35.6	43.9	53.2	63.3
0.0135	4.1	5.6	7.3	9.2	11.4	13.8	16.4	19.2	22.3	29.1	36.9	45.5	55.1	65.5
0.0140	4.2	5.8	7.5	9.5	11.8	14.2	16.9	19.9	23.1	30.1	38.1	47.1	57.0	67.8
0.0145	4.4	6.0	7.8	9.8	12.2	14.7	17.5	20.5	23.8	31.1	39.4	48.6	58.8	70.0
0.0150	4.5	6.1	8.0	10.2	12.5	15.2	18.1	21.2	24.6	32.1	40.6	50.2	60.7	72.3
0.0155	4.7	6.3	8.3	10.5	12.9	15.6	18.6	21.9	25.3	33.1	41.9	51.7	62.6	74.5
0.0160	4.8	6.5	8.5	10.8	13.3	16.1	19.2	22.5	26.1	34.1	43.1	53.3	64.4	76.7
0.0165	4.9	6.7	8.8	11.1	13.7	16.6	19.7	23.1	26.8	35.1	44.4	54.8	66.3	78.9
0.0170	5.1	6.9	9.0	11.4	14.1	17.0	20.3	23.8	27.6	36.0	45.6	56.3	68.1	81.1
0.0175	5.2	7.1	9.2	11.7	14.4	17.5	20.8	24.4	28.3	37.0	46.8	57.8	69.9	83.2
0.0180	5.3	7.3	9.5	12.0	14.8	17.9	21.3	25.1	29.1	37.9	48.0	59.2	71.7	85.4
0.0185	5.5	7.4	9.7	12.3	15.2	18.4	21.9	25.7	29.8	38.9	49.2	60.8	73.5	87.5
0.0190	5.6	7.6	10.0	12.6	15.6	18.8	22.4	26.3	30.5	39.9	50.4	62.3	75.3	89.7
0.0195	5.7	7.8	10.2	12.9	15.9	19.3	22.9	26.9	31.2	40.8	51.6	63.7	77.1	91.8
0.0200	5.9	8.0	10.4	13.2	16.3	19.7	23.5	27.5	31.9	41.7	52.8	65.2	78.9	93.9
0.0205	6.0	8.2	10.7	13.5	16.7	20.2	24.0	28.2	32.7	42.7	54.0	66.7	80.7	96.0
0.0210	6.1	8.3	10.9	13.8	17.0	20.6	24.5	28.8	33.4	43.6	55.2	68.1	82.4	98.1
0.0215	6.3	8.5	11.1	14.1	17.4	21.0	25.0	29.4	34.1	44.5	56.3	69.5	84.1	100.1
0.0220	6.4	8.7	11.4	14.4	17.7	21.5	25.6	30.0	34.8	45.4	57.5	71.0	85.9	102.2

TABLES B.2   *(Continued)*

Table B.2   *(40/5)*

$$f_v = 40\ 000 \text{ psi} \qquad f_c' = 5000 \text{ psi}$$

ρ	Effective Depth (in)													
	3.0	3.5	4.0	4.5	5.0	5.5	6.0	6.5	7.0	8.0	9.0	10.0	11.0	12.0
0.0225	6.5	8.9	11.6	14.7	18.1	21.9	26.1	30.6	35.5	46.3	58.6	72.4	87.6	104.3
0.0230	6.6	9.0	11.8	14.9	18.5	22.3	26.6	31.2	36.2	47.2	59.8	73.8	89.3	106.3
0.0235	6.8	9.2	12.0	15.2	18.8	22.8	27.1	31.8	36.9	48.1	60.9	75.2	91.0	108.3
0.0240	6.9	9.4	12.3	15.5	19.2	23.2	27.6	32.4	37.5	49.0	62.1	76.6	92.7	110.3
0.0245	7.0	9.6	12.5	15.8	19.5	23.6	28.1	33.0	38.2	49.9	63.2	78.0	94.4	112.3
0.0250	7.1	9.7	12.7	16.1	19.8	24.0	28.6	33.5	38.9	50.8	64.3	79.4	96.0	114.3
0.0255	7.3	9.9	12.9	16.4	20.2	24.4	29.1	34.1	39.6	51.7	65.4	80.8	97.7	116.3
0.0260	7.4	10.1	13.1	16.6	20.5	24.8	29.6	34.7	40.2	52.6	66.5	82.1	99.4	118.2
0.0265	7.5	10.2	13.4	16.9	20.9	25.2	30.0	35.3	40.9	53.4	67.6	83.5	101.0	120.2
0.0270	7.6	10.4	13.6	17.2	21.2	25.7	30.5	35.8	41.6	54.3	68.7	84.8	102.6	122.1
0.0275	7.8	10.6	13.8	17.4	21.5	26.1	31.0	36.4	42.2	55.1	69.8	86.1	104.2	124.1
0.0280	7.9	10.7	14.0	17.7	21.9	26.5	31.5	37.0	42.9	56.0	70.9	87.5	105.8	126.0
0.0285	8.0	10.9	14.2	18.0	22.2	26.9	32.0	37.5	43.5	56.8	71.9	88.8	107.4	127.9
0.0290	8.1	11.0	14.4	18.2	22.5	27.3	32.4	38.1	44.2	57.7	73.0	90.1	109.0	129.8
0.0295	8.2	11.2	14.6	18.5	22.9	27.7	32.9	38.6	44.8	58.5	74.0	91.4	119.6	131.6
0.0300	8.3	11.4	14.8	18.8	23.2	28.0	33.4	39.2	45.4	59.3	75.1	92.7	112.2	133.5
0.0305	8.5	11.5	15.0	19.0	23.5	28.4	33.8	39.7	46.1	60.2	76.1	94.0	113.7	135.4
0.0310	8.6	11.7	15.2	19.3	23.8	28.8	34.3	40.3	46.7	61.0	77.2	95.3	115.3	137.2
0.0315	8.7	11.8	15.4	19.5	24.1	29.2	34.8	40.8	47.3	61.8	78.2	96.5	116.8	139.0
0.0320	8.8	12.0	15.6	19.8	24.5	29.6	35.2	41.3	47.9	62.6	79.2	97.8	118.3	140.8
0.0325	8.9	12.1	15.8	20.1	24.8	30.0	35.7	41.8	48.5	63.4	80.2	99.1	119.9	142.6
0.0330	9.0	12.3	16.0	20.3	25.1	30.3	36.1	42.4	49.1	64.2	81.2	100.3	121.4	144.4
0.0335	9.1	12.4	16.2	20.6	25.4	30.7	36.6	42.9	49.8	65.0	82.2	101.5	122.9	146.2
0.0340	9.2	12.6	16.4	20.8	25.7	31.1	37.0	43.4	50.4	65.8	83.2	102.8	124.3	148.0
0.0345	9.4	12.7	16.6	21.1	26.0	31.5	37.4	43.9	50.9	66.5	84.2	104.0	125.8	149.7
0.0350	9.5	12.9	16.8	21.3	26.3	31.8	37.9	44.4	51.5	67.3	85.2	105.2	127.3	151.5
0.0355	9.6	13.0	17.0	21.5	26.6	32.2	38.3	44.9	52.1	68.1	86.2	106.4	128.7	153.2
0.0360	9.7	13.2	17.2	21.8	26.9	32.5	38.7	45.5	52.7	68.9	87.1	107.6	130.2	154.9
0.0365	9.8	13.3	17.4	22.0	27.2	32.9	39.2	46.0	53.3	69.6	88.1	108.8	131.6	156.6
0.0370	9.9	13.5	17.6	22.3	27.5	33.3	39.6	46.4	53.9	70.4	89.0	109.9	133.0	158.3
0.0375	10.0	13.6	17.8	22.5	27.8	33.6	40.0	46.9	54.4	71.1	90.0	111.1	134.4	160.0
0.0380	10.1	13.8	18.0	22.7	28.1	34.0	40.4	47.4	55-0	71.8	90.9	112.3	135.8	161.7
0.0385	10.2	13.9	18.1	23.0	28.4	34.3	40.8	47.9	55.6	72.6	91.9	113.4	137.2	163.3
0.0390	10.3	14.0	18.3	23.2	28.6	34.7	41.2	48.4	56.1	73.3	92.8	114.6	138.6	165.0
0.0395	10.4	14.2	18.5	23.4	28.9	35.0	41.6	48.9	56.7	74.0	93.7	115.7	140.0	166.6
0.0400	10.5	14.3	18.7	23.7	29.2	35.3	42.1	49.4	57.2	74.8	94.6	116.8	141.3	168.2
0.0405	10.6	14.4	18.9	23.9	29.5	35.7	42.5	49.8	57.8	75.5	95.5	117.9	142.7	169.8
0.0410	10.7	14.6	19.0	24.1	29.8	36.0	42.9	50.3	58.3	76.2	96.4	119.0	144.0	171.4
0.0415	10.8	14.7	19.2	24.3	30.0	36.3	43.2	50.8	58.9	76.9	97.3	120.1	145.4	173.0
0.0420	10.9	14.9	19.4	24.5	30.3	36.7	43.6	51.2	59.4	77.6	98.2	121.2	146.7	174.6
0.0425	11.0	15.0	19.6	24.8	30.6	37.0	44.0	51.7	59.9	78.3	99.1	122.3	148.0	176.1
0.0430	11.1	15.1	19.7	25.0	30.8	37.3	44.4	52.1	60.5	79.0	99.9	123.4	149.3	177.7
0.0435	11.2	15.2	19.9	25.2	31.1	37.6	44.8	52.6	61.0	79.6	100.8	124.4	150.6	179.2

TABLES B.2 M_r (kip-ft) for Sections 12 in wide

Table B.2 (60/3)

$$f_y = 60\ 000 \text{ psi} \qquad f'_c = 3000 \text{ psi}$$

ρ	3.0	3.5	4.0	4.5	5.0	5.5	6.0	6.5	7.0	8.0	9.0	10.0	11.0	12.0
						Effective Depth (in)								
0.0020	0.9	1.3	1.7	2.1	2.6	3.2	3.8	4.5	5.2	6.7	8.5	10.5	12.8	15.2
0.0025	1.2	1.6	2.1	2.7	3.3	4.0	4.7	5.5	6.4	8.4	10.6	13.1	15.9	18.9
0.0030	1.4	1.9	2.5	3.2	3.9	4.7	5.6	6.6	7.7	10.0	12.7	15.6	18.9	22.5
0.0035	1.6	2.2	2.9	3.7	4.5	5.5	6.5	7.7	8.9	11.6	14.7	18.1	21.9	26.1
0.0040	1.9	2.5	3.3	4.2	5.1	6.2	7.4	8.7	10.1	13.2	16.7	20.6	24.9	29.6
0.0045	2.1	2.8	3.7	4.7	5.8	7.0	8.3	9.7	11.3	14.7	18.6	23.0	27.8	33.1
0.0050	2.3	3.1	4.1	5.1	6.4	7.7	9.1	10.7	12.4	16.3	20.6	25.4	30.7	36.6
0.0055	2.5	3.4	4.4	5.6	6.9	8.4	10.0	11.7	13.6	17.8	22.5	27.8	33.6	40.0
0.0060	2.7	3.7	4.8	6.1	7.5	9.1	10.8	12.7	14.8	19.3	24.4	30.1	36.4	43.4
0.0065	2.9	4.0	5.2	6.6	8.1	9.8	11.7	13.7	15.9	20.7	26.3	32.4	39.2	46.7
0.0070	3.1	4.2	5.5	7.0	8.7	10.5	12.5	14.7	17.0	22.2	28.1	34.7	42.0	49.9
0.0075	3.3	4.5	5.9	7.5	9.2	11.2	13.3	15.6	18.1	23.6	29.9	36.9	44.7	53.2
0.0080	3.5	4.8	6.3	7.9	9.8	11.8	14.1	16.5	19.2	25.0	31.7	39.1	47.3	56.3
0.0085	3.7	5.1	6.6	8.4	10.3	12.5	14.9	17.4	20.2	26.4	33.4	41.3	50.0	59.5
0.0090	3.9	5.3	7.0	8.8	10.9	13.1	15.6	18.4	21.3	27.8	35.2	43.4	52.6	62.6
0.0095	4.1	5.6	7.3	9.2	11.4	13.8	16.4	19.2	22.3	29.2	36.9	45.5	55.1	65.6
0.0100	4.3	5.8	7.6	9.6	11.4	14.4	17.1	20.1	22.3	30.5	38.6	47.6	57.6	68.6
0.0105	4.5	6.1	7.9	10.1	11.4	15.0	17.9	21.0	22.3	31.8	40.2	49.7	60.1	71.5
0.0110	4.7	6.3	8.3	10.5	11.4	15.6	18.6	21.8	22.3	33.1	41.9	51.7	62.5	74.4
0.0115	4.8	6.6	8.6	10.9	11.4	16.2	19.3	22.7	22.3	34.4	43.5	53.7	64.9	77.3
0.0120	5.0	6.8	8.9	11.3	11.4	16.8	20.0	23.5	22.3	35.6	45.1	55.6	67.3	80.1
0.0125	5.2	7.0	9.2	11.7	11.4	17.4	20.7	24.3	28.2	36.8	46.6	57.5	69.6	82.9
0.0130	5.3	7.3	9.5	12.0	11.4	18.0	21.4	25.1	29.1	38.0	46.6	59.4	71.9	85.6
0.0135	5.5	7.5	9.8	12.4	15.3	18.5	22.1	25.9	30.0	39.2	46.6	61.3	74.2	88.3
0.0140	5.7	7.7	10.1	12.8	15.8	19.1	22.7	26.7	30.9	40.4	46.6	63.1	76.4	90.9
0.0145	5.8	8.0	10.4	13.1	16.2	19.6	23.4	27.4	31.8	41.5	46.6	64.9	78.5	93.5
0.0150	6.0	8.2	10.7	13.5	16.7	20.2	24.0	28.2	32.7	42.7	54.0	66.7	80.7	96.0
0.0155	6.2	8.4	10.9	13.8	17.1	20.7	24.6	28.9	33.5	43.8	55.4	68.4	82.8	98.5
0.0160	6.3	8.6	11.2	14.2	17.5	21.2	25.2	29.6	34.3	44.9	56.8	70.1	84.8	100.9

TABLES B.2   *M_r (kip-ft) for Sections 12 in wide*

Table B.2   *(60/4)*

f_y = 60 000 psi f'_c = 4000 psi

Effective Depth (in)

ρ	3.0	3.5	4.0	4.5	5.0	5.5	6.0	6.5	7.0	8.0	9.0	10.0	11.0	12.0
0.0020	1.0	1.3	1.7	2.1	2.7	3.2	3.8	4.5	5.2	6.8	8.6	10.6	12.8	15.3
0.0025	1.2	1.6	2.1	2.7	3.3	4.0	4.8	5.6	6.5	8.4	10.7	13.2	16.0	19.0
0.0030	1.4	1.9	2.5	3.2	3.9	4.8	5.7	6.7	7.7	10.1	12.8	15.8	19.1	22.7
0.0035	1.6	2.2	2.9	3.7	4.6	5.5	6.6	7.7	9.0	11.7	14.8	18.3	22.2	26.4
0.0040	1.9	2.6	3.3	4.2	5.2	6.3	7.5	8.8	10.2	13.3	16.9	20.8	25.2	30.0
0.0045	2.1	2.9	3.7	4.7	5.8	7.1	8.4	9.9	11.4	14.9	18.9	23.3	28.2	33.6
0.0050	2.3	3.2	4.1	5.2	6.5	7.8	9.3	19.9	12.6	16.5	20.9	25.8	31.2	37.2
0.0055	2.5	3.5	4.5	5.7	7.1	8.5	10.2	11.9	13.8	18.1	22.9	28.3	34.2	40.7
0.0060	2.8	3.8	4.9	6.2	7.7	9.3	11.0	13.0	15.0	19.6	24.9	30.7	37.1	44.2
0.0065	3.0.	4.1	5.3	6.7	8.3	10.0	11.9	14.0	16.2	21.2	26.8	33.1	40.0	47.6
0.0070	3.2	4.3	5.7	7.2	8.9	10.7	12.8	15.0	17.4	22.7	28.7	35.5	42.9	51.1
0.0075	3.4	4.6	6.0	7.7	9.5	11.4	13.6	16.0	18.5	24.2	30.6	37.8	45.8	54.4
0.0080	3.6	4.9	6.4	8.1	10.0	12.1	14.5	17.0	19.7	25.7	32.5	40.1	48.6	57.8
0.0085	3.8	5.2	6.8	8.6	10.6	12.8	15.3	17.9	20.8	27.2	34.4	42.4	51.4	61.1
0.0090	4.0	5.5	7.2	9.1	11.2	13.5	16.1	18.9	21.9	28.6	36.2	44.7	54.1	64.4
0.0095	4.2	5.8	7.5	9.5	11.7	14.2	16.9	19.9	23.0	30.1	38.1	47.0	56.9	67.7
0.0100	4.4	6.0	7.9	10.0	12.3	14.9	17.7	20.8	24.1	31.5	39.9	49.2	59.6	70.9
0.0105	4.6	6.3	8.2	10.4	12.9	15.6	18.5	21.7	25.2	32.9	41.7	51.4	62.2	74.1
0.0110	4.8	6.6	8.6	10.9	13.4	16.2	19.3	22.7	26.3	34.3	43.4	53.6	64.9	77.2
0.0115	5.0	6.8	8.9	11.3	13.9	16.9	20.1	23.6	27.3	35.7	45.2	55.8	67.5	80.3
0.0120	5.2	7.1	9.3	11.7	14.5	17.5	20.9	24.5	28.4	37.1	46.9	57.9	70.1	83.4
0.0125	5.4	7.4	9.6	12.2	15.0	18.2	21.6	25.4	29.4	38.4	48.6	60.0	72.6	86.4
0.0130	5.6	7.6	9.9	12.6	15.5	18.8	22.4	26.2	30.4	39.8	50.3	62.1	75.2	89.5
0.0135	5.8	7.9	10.3	13.0	16.0	19.4	23.1	27.1	31.5	41.1	52.0	64.2	77.7	92.4
0.0140	6.0	8.1	10.6	13.4	16.6	20.0	23.8	28.0	32.5	42.4	53.6	66.2	80.1	95.4
0.0145	6.1	8.4	10.9	13.8	17.1	20.6	24.6	28.8	33.4	43.7	55.3	68.3	82.6	98.3
0.0150	6.3	8.6	11.2	14.2	17.6	21.2	25.3	29.7	34.4	45.0	56.9	70.2	85.0	101.2
0.0155	6.5	8.8	11.6	14.6	18.1	21.8	26.0	30.5	35.4	46.2	58.5	72.2	87.4	104.0
0.0160	6.7	9.1	11.9	15.0	18.5	22.4	26.7	31.3	36.3	47.5	60.1	74.2	89.7	106.8
0.0165	6.8	9.3	12.2	15.4	19.0	23.0	27.4	32.1	37.3	48.7	61.6	76.1	92.1	109.6
0.0170	7.0	9.6	12.5	15.8	19.5	23.6	28.1	33.0	38.2	49.9	63.2	78.0	94.4	112.3
0.0175	7.2	9.8	12.8	16.2	20.0	24.2	28.8	33.7	39.1	51.1	64.7	79.9	96.6	115.0
0.0180	7.4	10.0	13.1	16.5	20.4	24.7	29.4	34.5	40.0	52.3	66.2	81.7	98.9	117.7
0.0185	7.5	10.2	13.4	16.9	20.9	25.3	30.1	35.3	40.9	53.5	67.7	83.5	101.1	120.3
0.0190	7.7	10.5	13.7	17.3	21.3	25.8	30.7	36.1	41.8	54.6	69.1	85.3	103.3	122.9
0.0195	7.8	10.7	13.9	17.6	21.8	26.4	31.4	36.8	42.7	55.8	70.6	87.1	105.4	125.5
0.0200	8.0	10.9	14.2	18.0	22.2	26.9	32.0	37.6	43.6	56.9	72.0	88.9	107.5	128.0
0.0205	8.2	11.1	14.5	18.3	22.7	27.4	32.6	38.3	44.4	58.0	73.4	90.6	109.6	130.5
0.0210	8.3	11.3	14.8	18.7	23.1	27.9	33.2	39.0	45.2	59.1	74.8	92.3	111.7	132.9

TABLES B.2   *M$_r$ (kip-ft) for Sections 12 in wide*

Table B.2   (60/5)

$$f_v = 60\ 000 \text{ psi} \qquad f_c' = 5000 \text{ psi}$$

Effective Depth (in)

ρ	3.0	3.5	4.0	4.5	5.0	5.5	6.0	6.5	7.0	8.0	9.0	10.0	11.0	12.0
0.0020	1.0	1.3	1.7	2.2	2.7	3.2	3.8	4.5	5.2	6.8	8.6	10.6	12.9	15.3
0.0025	1.2	1.6	2.1	2.7	3.3	4.0	4.8	5.6	6.5	8.5	10.7	13.3	16.0	19.1
0.0030	1.4	1.9	2.5	3.2	4.0	4.8	5.7	6.7	7.8	10.1	12.8	15.9	19.2	22.8
0.0035	1.7	2.3	2.9	3.7	4.6	5.6	6.6	7.8	9.0	11.8	14.9	18.4	22.3	26.5
0.0040	1.9	2.6	3.4	4.3	5.2	6.3	7.6	8.9	10.3	13.4	17.0	21.0	25.4	30.2
0.0045	2.1	2.9	3.8	4.8	5.9	7.1	8.5	9.9	11.5	15.1	19.1	23.5	28.5	33.9
0.0050	2.3	3.2	4.2	5.3	6.5	7.9	9.4	11.0	12.8	16.7	21.1	26.0	31.5	37.5
0.0055	2.6	3.5	4.6	5.8	7.1	8.6	10.3	12.1	14.0	18.3	23.1	28.5	34.5	41.1
0.0060	2.8	3.8	5.0	6.3	7.8	9.4	11.2	13.1	15.2	19.9	25.1	31.0	37.5	44.7
0.0065	3.0	4.1	5.4	6.8	8.4	10.1	12.1	14.1	16.4	21.4	27.1	33.5	40.5	48.2
0.0070	3.2	4.4	5.7	7.3	9.0	10.9	12.9	15.2	17.6	23.0	29.1	35.9	43.5	51.7
0.0075	3.5	4.7	6.1	7.8	9.6	11.6	13.8	16.2	18.8	24.5	31.1	38.3	46.4	55.2
0.0080	3.7	5.0	6.5	8.3	10.2	12.3	14.7	17.2	20.0	26.1	33.0	40.8	49.3	58.7
0.0085	3.9	5.3	6.9	8.7	10.8	13.0	15.5	18.2	21.1	27.6	34.9	43.1	52.2	62.1
0.0090	4.1	5.6	7.3	9.2	11.4	13.8	16.4	19.2	22.3	29.1	36.9	45.5	55.1	65.5
0.0095	4.3	5.9	7.7	9.7	12.0	14.5	17.2	20.2	23.4	30.6	38.8	47.8	57.9	68.9
0.0100	4.5	6.1	8.0	10.2	12.5	15.2	18.1	21.2	24.6	32.1	40.6	50.2	60.7	72.3
0.0105	4.7	6.4	8.4	10.6	13.1	15.9	18.9	22.2	25.7	33.6	42.5	52.5	63.5	75.6
0.0110	4.9	6.7	8.8	11.1	13.7	16.6	19.7	23.1	26.8	35.1	44.4	54.8	66.3	78.9
0.0115	5.1	7.0	9.1	11.6	14.3	17.3	20.5	24.1	28.0	36.5	46.2	57.0	69.0	82.1
0.0120	5.3	7.3	9.5	12.0	14.8	17.9	21.3	25.1	29.1	37.9	48.0	59.3	71.7	85.4
0.0125	5.5	7.5	9.8	12.5	15.4	18.6	22.1	26.0	30.1	39.4	49.8	61.5	74.4	88.6
0.0130	5.7	7.8	10.2	12.9	15.9	19.3	22.9	26.9	31.2	40.8	51.6	63.7	77.1	91.8
0.0135	5.9	8.1	10.5	13.4	16.5	19.9	23.7	27.9	32.3	42.2	53.4	65.9	79.8	94.9
0.0140	6.1	8.3	10.9	13.8	17.0	20.6	24.5	28.8	33.4	43.6	55.2	68.1	82.4	98.1
0.0145	6.3	8.6	11.2	14.2	17.6	21.3	25.3	29.7	34.4	45.0	56.9	70.3	85.0	101.2
0.0150	6.5	8.9	11.6	14.7	18.1	21.9	26.1	30.6	35.5	46.3	58.6	72.4	87.6	104.3
0.0155	6.7	9.1	11.9	15.1	18.6	22.5	26.8	31.5	36.5	47.7	60.4	74.5	90.2	107.3
0.0160	6.9	9.4	12.3	15.5	19.2	23.2	27.6	32.4	37.5	49.0	62.1	76.6	92.7	110.3
0.0165	7.1	9.6	12.6	15.9	19.7	23.8	28.3	33.2	38.6	50.4	63.7	78.7	95.2	113.3
0.0170	7.3	9.9	12.9	16.4	20.2	24.4	29.1	34.1	39.6	51.7	65.4	80.8	97.7	116.3
0.0175	7.5	10.1	13.2	16.8	20.7	25.0	29.8	35.0	40.6	53.0	67.1	82.8	100.2	119.2
0.0180	7.6	10.4	13.6	17.2	21.2	25.7	30.5	35.8	41.6	54.3	68.7	84.8	102.6	122.1
0.0185	7.8	10.6	13.9	17.6	21.7	26.3	31.3	36.7	42.5	55.6	70.3	86.8	105.0	125.0
0.0190	8.0	10.9	14.2	18.0	22.2	26.9	32.0	37.5	43.5	56.8	71.9	88.8	107.4	127.9
0.0195	8.2	11.1	14.5	18.4	22.7	27.5	32.7	38.3	44.5	58.1	73.5	90.8	109.8	130.7
0.0200	8.3	11.4	14.8	18.8	23.2	28.0	33.4	39.2	45.4	59.3	75.1	92.7	112.2	133.5
0.0205	8.5	11.6	15.1	19.2	23.7	28.6	34.1	40.0	46.4	60.6	76.7	94.6	114.5	136.3
0.0210	8.7	11.8	15.4	19.5	24.1	29.2	34.8	40.8	47.3	61.8	78.2	96.5	116.8	139.0
0.0215	8.9	12.1	15.7	19.9	24.6	29.8	35.4	41.6	48.2	63.0	79.7	98.4	119.1	141.7
0.0220	9.0	12.3	16.0	20.3	25.1	30.3	36.1	42.4	49.1	64.2	81.2	100.3	121.4	144.4
0.0225	9.2	12.5	16.3	20.7	25.5	30.9	36.8	43.2	50.1	65.4	82.7	102.1	123.6	147.1
0.0230	9.4	12.7	16.6	21.1	26.0	31.5	37.4	43.9	50.9	66.5	84.2	104.0	125.8	149.7
0.0235	9.5	13.0	16.9	21.4	26.4	32.0	38.1	44.7	51.8	67.7	85.7	105.8	128.0	152.3
0.0240	9.7	13.2	17.2	21.8	26.9	32.5	38.7	45.5	52.7	68.9	87.1	107.6	130.2	154.9
0.0245	9.8	13.4	17.5	22.1	27.3	33.1	39.4	46.2	53.6	70.0	88.6	109.4	132.3	157.5
0.0250	10.0	13.6	17.8	22.5	27.8	33.6	40.0	46.9	54.4	71.1	90.0	111.1	134.4	160.0

GRAPHS C.1 *Effective Depths Required for Various M_r Values at Selected Widths*

Graph C.1 (3)

$$\rho = \tfrac{2}{3}\,\rho_{max}$$

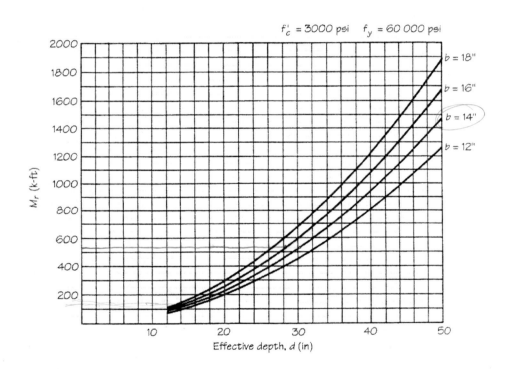

GRAPHS C.1   *Effective Depths Required for Various M_r Values at Selected Widths*

Graph C.1 (4)

$$\rho = \tfrac{2}{3}\,\rho_{max}$$

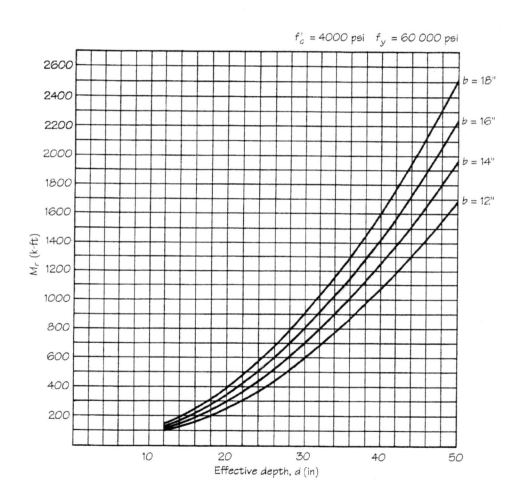

GRAPHS C.1 *Effective Depths Required for Various M_r Values at Selected Widths*

Graph C.1 (5)

$$\rho = \tfrac{2}{3}\, \rho_{max}$$

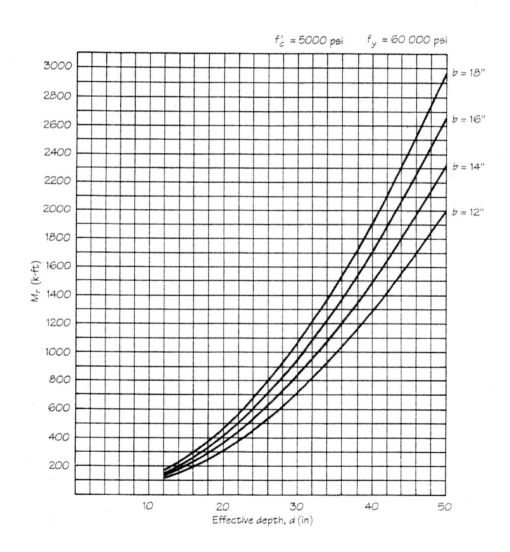

GRAPHS C.2   Strength Design Capacities for Preliminary Design of Selected Laterally Tied Columns

Graph C.2 (2%)

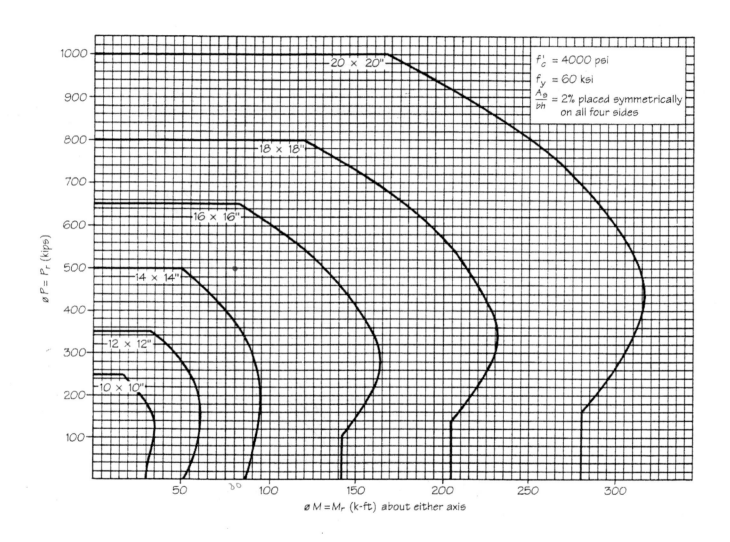

GRAPHS C.2 *Strength Design Capacities for Preliminary Design of Selected Laterally Tied Columns*

Graph C.2 (4%)

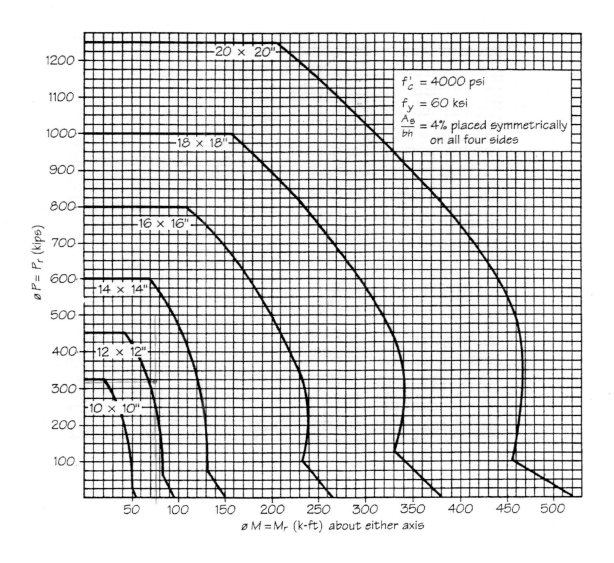

GRAPHS C.2   *Strength Design Capacities for Preliminary Design of Selected Laterally Tied Columns*

Graph C.2 (6%)

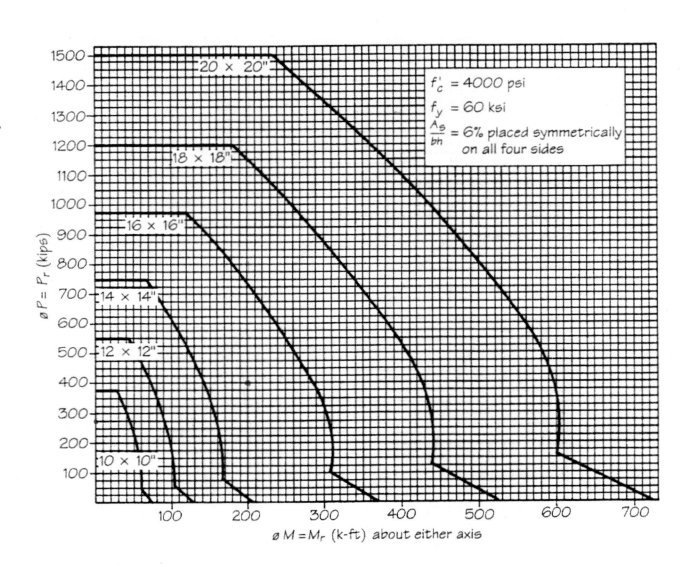

$\phi M = M_r$ (k-ft) about either axis

TABLE D   *Shear, Moment, and Deflection Formulas*

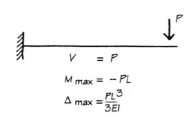

$$V = P$$
$$M_{max} = -PL$$
$$\Delta_{max} = \frac{PL^3}{3EI}$$

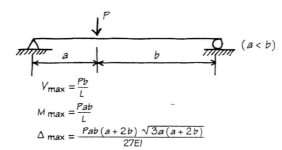

$(a < b)$

$$V_{max} = \frac{Pb}{L}$$
$$M_{max} = \frac{Pab}{L}$$
$$\Delta_{max} = \frac{Pab(a+2b)\sqrt{3a(a+2b)}}{27EI}$$

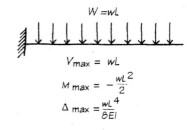

$$V_{max} = wL$$
$$M_{max} = -\frac{wL^2}{2}$$
$$\Delta_{max} = \frac{wL^4}{8EI}$$

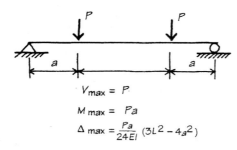

$$V_{max} = P$$
$$M_{max} = Pa$$
$$\Delta_{max} = \frac{Pa}{24EI}(3L^2 - 4a^2)$$

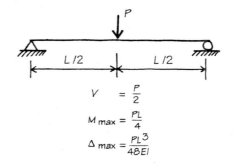

$$V = \frac{P}{2}$$
$$M_{max} = \frac{PL}{4}$$
$$\Delta_{max} = \frac{PL^3}{48EI}$$

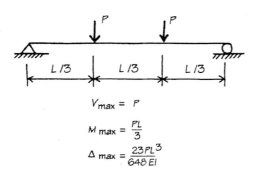

$$V_{max} = P$$
$$M_{max} = \frac{PL}{3}$$
$$\Delta_{max} = \frac{23PL^3}{648EI}$$

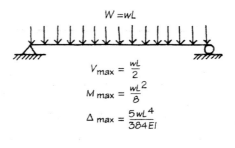

$W = wL$

$$V_{max} = \frac{wL}{2}$$
$$M_{max} = \frac{wL^2}{8}$$
$$\Delta_{max} = \frac{5wL^4}{384EI}$$

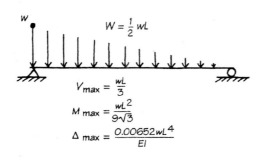

$W = \frac{1}{2}wL$

$$V_{max} = \frac{wL}{3}$$
$$M_{max} = \frac{wL^2}{9\sqrt{3}}$$
$$\Delta_{max} = \frac{0.00652wL^4}{EI}$$

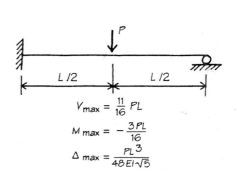

$$V_{max} = \frac{11}{16}PL$$
$$M_{max} = -\frac{3PL}{16}$$
$$\Delta_{max} = \frac{PL^3}{48EI\sqrt{5}}$$

(continued)

TABLE D   *(Continued)*

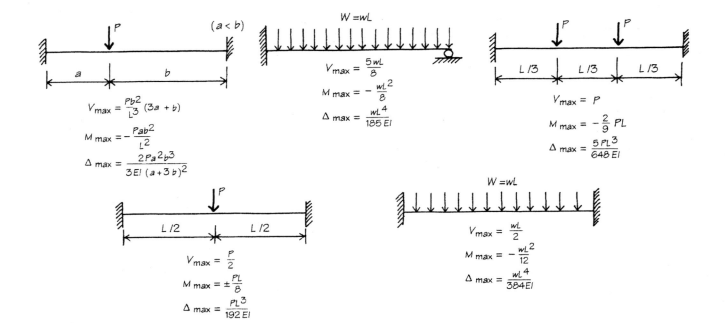

$V_{max} = \dfrac{Pb^2}{L^3}(3a + b)$

$M_{max} = -\dfrac{Pab^2}{L^2}$

$\Delta_{max} = \dfrac{2Pa^2b^3}{3EI\,(a+3b)^2}$

$V_{max} = \dfrac{5wL}{8}$

$M_{max} = -\dfrac{wL^2}{8}$

$\Delta_{max} = \dfrac{wL^4}{185\,EI}$

$V_{max} = P$

$M_{max} = -\dfrac{2}{9}PL$

$\Delta_{max} = \dfrac{5PL^3}{648\,EI}$

$V_{max} = \dfrac{P}{2}$

$M_{max} = \pm\dfrac{PL}{8}$

$\Delta_{max} = \dfrac{PL^3}{192\,EI}$

$V_{max} = \dfrac{wL}{2}$

$M_{max} = -\dfrac{wL^2}{12}$

$\Delta_{max} = \dfrac{wL^4}{384\,EI}$

ANSWERS TO PROBLEMS

Chapter 6

6.1 No, 425 kip-ft > 366 kip-ft

6.2 572 kip-ft

6.3 2.1 kips/ft

6.4 More, 862 kip-ft > 801 kip-ft

6.5 218 kips

6.6 No, the section violates Code.

6.7 104 kips

Chapter 7

7.1 (*a*) 27 in
(*b*) 20 in

7.2 0.0071

7.3 29.5 in

7.4 22 in

7.5 With an assumed d of 26 in, a new ρ of 0.0096 will be required.

7.6 36 in (estimated $h = 40$ in)

7.7 27 in

7.8 22 in

7.9 32.5 in (estimated $h = 38$ in)

7.10 32 in (estimated $h = 37$ in)

7.11 36.5 in (estimated $h = 42$ in)

7.12 38 in (estimated $h = 43$ in)

7.13 See Figure A7.13.

7.14 See Figure A7.14.

Chapter 8

8.1 1/3, 5/7, 1/10, 6/12 EE

(Note: To avoid having just one stirrup at a certain spacing, a better placement might be 1/3, 7/7, 6/12 EE)

8.2 1/4, 8/8, 8/15 EE

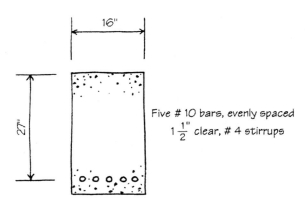

FIGURE A7.13

Five # 10 bars, evenly spaced
$1\frac{1}{2}''$ clear, # 4 stirrups

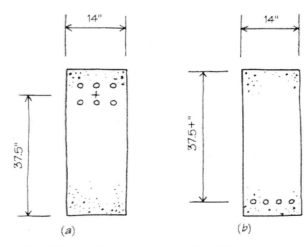

(a)

Six #9 bars in two layers, evenly spaced

(b)

Four #9 bars, evenly spaced
(Note: with only one layer of steel, d can be larger and still maintain the required cover.)

FIGURE A7.14

8.3 69.5 in (assumed $h = 21$ in)

8.4 From the support: 1/3, 8/6

8.5 (Assumed h = 28 in.)
From the left end: 1/6, 18/12
From the right end: 1/2, 8/5, 3/8, 4/12

8.6 1/2, 6/4, 11/11 EE

Chapter 9

9.1 (*a*) 0.233 in
(*b*) 0.699 in

9.2 Yes, 0.953 in < 1.33 in

9.3 $w = 3.16$ kips/ft

9.4 No, 2.18 in > 2.0 in

Chapter 10

10.1 No, from Graph C.2 (2%)

10.2 Yes, from Graph C.2 (4%)

10.3 See Figure A10.3.

10.4 14 × 14-in section from Graph C.2 (6%)

10.5 Yes, from Graph C.2 (4%)

10.6 14 × 14-in section; top floor: $\rho_g = 6\%$;
first floor: $\rho_g = 2\%$

Chapter 11

11.1 Moment steel: #4 @ 15.5 in; *t/s*: #3 @ 8.5 in

11.2 103 psf

11.3 Moment steel: #5 @ 17 in; *t/s*: #4 @ 13 in

11.4 Moment steel: #4 @ 11 in; *t/s*: #3 @ 6 in

11.5 See Figure A11.5.

Chapter 13

13.1 Area okay: 49 ft² > 48.8 ft²; punching shear okay: 223 kips > 203 kips; A_s okay: 1.97 in² < 3.09 in²

FIGURE A10.3

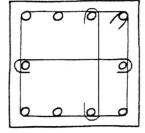

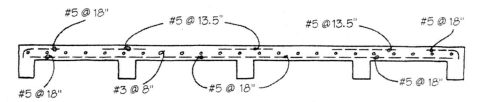

FIGURE A11.5

13.2 No, 195 kips < 233 kips

13.3 h = 10 in, d=6.5in
B = 4 ft-6 in
moment steel: #5 @ 11.5 in
longitudinal steel: six #4

13.4 Bearing okay: 3030 psf > 1850 psf; punching shear okay: 248 kips > 231 kips; beam shear not okay: 89.7 kips < 91.0 kips; A_s in long direction not okay: 6.00 in² < 6.22 in²; A_s in short direction not okay: s > 18 in

13.5 h = 12 in, d = 8 in
B = 3 ft-4 in
moment steel: #6 @ 11 in
longitudinal steel: five #4

Chapter 14

14.1 (*a*) Sliding not okay: needs 24.5-in shear key

(*b*) Overturning okay: F.S. = 3.77
Soil bearing capacity okay: 3500 psf > 2090 psf

14.2 #6 @ 7.5 in

14.3 5 ft-4 in

Chapter 15

15.1 f_t = − 2.89 ksi
f_b = +0.03 ksi

15.2 18.7 in

15.3 3.56 kips/ft

15.4 (*a*) 90.4 kips
(*b*) 29.2 ft

15.5 (*a*) 12 in
(*b*) Yes

INDEX

LIST OF SYMBOLS USED

a	depth of fictitious stress block
a_s	cross-sectional area of one bar
A_c	area of concrete (transformed tension steel)
A'_c	area of concrete (transformed compression steel)
A_r	area required (footing)
A_s	cross-sectional area of total tension steel
A'_s	cross-sectional area of total compression steel
A_v	cross-sectional area of one stirrup
b	width
b_o	perimeter (punching shear)
B	width of base (retaining wall)
B_r	width required (wall footing)
c	distance from neutral axis to extreme bending fiber
C	compressive force resultant
C_a	active lateral pressure coefficient (soil)
C_p	passive lateral pressure coefficient (soil)
d	depth or effective depth
D	unfactored dead load or its effect
e	eccentricity
E_c	modulus of elasticity of concrete
E_s	modulus of elasticity of steel
EE	each end (stirrup placement)
f	stress
f_b	bending stress
f_c	concrete stress
f'_c	28-day compressive strength of concrete (cylinder test)
f'_{ci}	compressive strength of concrete at the time of prestressing
f_p	stress due to prestress effects
f_r	coefficient of friction (soil/concrete)
f_y	yield stress
F_b	allowable bending stress
F_r	force due to friction

F.S.	factor of safety
h	height, also thickness of slab or column, overall depth of beam
h_k	depth of shear key
I	moment of inertia
I_{cr}	cracked section moment of inertia
I_e	effective moment of inertia
I_g	gross section moment of inertia
k	kip(s)
kcf	kips per cubic foot
klf	kips per linear foot
ksi	kips per square inch
L	length of member, also span, also unfactored live load or its effect
lb	pound(s)
M	moment
M_o	overturning moment
M_r	resisting moment after application of ϕ
M_u	ultimate moment due to factored loads
n	modular ratio
N	summation of vertical forces (retaining wall)
n.a.	neutral axis
p	soil pressure
p_a	active soil pressure
p_p	passive soil pressure
pcf	pounds per cubic foot
plf	pounds per linear foot
psf	pounds per square foot
psi	pounds per square inch
P	concentrated load, also prestress force
P_{dl}	concentrated unfactored dead load
P_i	initial prestress force
P_k	resisting force provided by shear key
P_{ll}	concentrated unfactored live load
P_r	ultimate axial load capacity (column) after application of ϕ

P_u	ultimate factored concentrated load
q	allowable soil bearing pressure
q_e	effective allowable soil bearing pressure
q_u	soil bearing pressure due to factored leads
R	coefficient of resistance (beams)
s	spacing of reinforcing bars or stirrups
s_{max}	maximum permissible spacing of stirrups or reinforcing bars in slabs
S	elastic section modulus
s.w.	self-weight
T	tensile force resultant
U	ultimate factored load or its effect
V	shearing force
V_c	shear capacity of concrete
V_s	shear capacity of steel
V_u	ultimate shear force due to factored loads
w	load per unit length, also unit weight
w_u	ultimate factored load per unit length
w_{dl}	distributed unfactored dead load
w_{ll}	distributed unfactored live load
W	total distributed load
w/c	water-to-cement ratio by weight
y	centroidal distance
Δ	deflection
$\Delta_{i_{dl}}$	initial deflection due to dead load
$\Delta_{i_{ll}}$	initial deflection due to live load
$\Delta_{i_{sl}}$	initial deflection due to sustained load
Δ_{lt}	long-term deflection
$\Delta_{rem_{ll}}$	deflection due to remaining live load
Δ_t	total deflection
ø	capacity reduction factor, also internal angle of friction (soil)

ρ	tensile steel reinforcement ratio
ρ'	compressive steel reinforcement ratio
ρ_b	balanced steel reinforcement ratio
ρ_g	gross steel reinforcement ratio (column)
ρ_{min}	minimum steel reinforcement ratio
ρ_{max}	maximum steel reinforcement ratio
ρ_t	temperature/shrinkage steel reinforcement ratio